WAR AND THE ENVIRONMENT

Number 125: Williams-Ford Texas A&M University
Military History Series

WAR AND THE ENVIRONMENT

Military Destruction in the Modern Age

Edited by Charles E. Closmann

Texas A&M University Press
College Station

Copyright © 2009 Charles E. Closmann
Manufactured in the United States of America
All rights reserved
First edition

This paper meets the requirements of ANSI/NISO Z39.48-1992
(Permanence of Paper).
Binding materials have been chosen for durability.

Library of Congress Cataloging-in-Publication Data

War and the environment : military destruction in the modern age / edited by
Charles E. Closmann.—1st ed.
p. cm.—(Williams-Ford Texas A&M University military history series ; no. 125)
Includes index.
ISBN-13: 978-1-60344-115-5 (cloth : alk. paper)
ISBN-10: 1-60344-115-8 (cloth : alk. paper)
ISBN-13: 978-1-60344-169-8 (pbk. : alk. paper)
ISBN-10: 1-60344-169-7 (pbk. : alk. paper)
1. War—Environmental aspects—History. 2. Armed Forces—Environmental aspects—History. 3. Postwar reconstruction—Environmental aspects—History. I. Closmann, Charles E. (Charles Edwin), 1957– II. Series: Williams-Ford Texas A&M University military history series ; no. 125.
TD195.W29W37 2009
363.7—dc22
2008051134

CONTENTS

List of Tables, Maps, and Illustrations / **vii**

Acknowledgments / **ix**

Introduction: Landscapes of Peace, Environments of War / **1**
Charles E. Closmann

Chapter 1. The Global Environmental Footprint of the U.S. Military, 1789–2003 / **10**
J. R. McNeill and David S. Painter

Chapter 2. Wood for War: The Legacy of Human Conflict on the Forests of the Philippines, 1565–1946 / **32**
Greg Bankoff

Chapter 3. Devouring the Land: Sherman's 1864–1865 Campaigns / **49**
Lisa M. Brady

Chapter 4. Environments of Death: Trench Warfare on the Western Front, 1914–1918 / **68**
Dorothee Brantz

Chapter 5. Total War? Administering Germany's Environment in Two World Wars / **92**
Frank Uekötter

Chapter 6. World War II and the Axis of Disease: Battling Malaria in Twentieth-Century Italy / **112**
Marcus Hall

CHAPTER 7. Birds on the Home Front: Wildlife Conservation in the Western United States during World War II / **132**
Robert Wilson

CHAPTER 8. Creating the Natural Fortress: Landscape, Resistance, and Memory in the Vercors, France / **150**
Chris Pearson

CHAPTER 9. Wartime Destruction and the Postwar Cityscape / **171**
Jeffry M. Diefendorf

CONTRIBUTORS / **193**

INDEX / **196**

LIST OF ILLUSTRATIONS

TABLES
1.1. Manpower of U.S. military on active duty, 1789–2000 / **12**
1.2. Energy consumption of selected U.S. military equipment / **24**
2.1. Forest cover and population, 1565–1948 / **35**

MAPS
3.1. Sherman's march through Georgia and the Carolinas, 1864–65 / **56**
6.1. Italy, Sardinia, and Corsica / **113**
7.1. National wildlife refuges in California, 1955 / **136**

FIGURES AND ILLUSTRATIONS
2.1. Timber delivered to the *Colonia Militar* at Tumauini / **40**
3.1. Engraving of Sherman's troops, 1868 / **53**
4.1. Battlefield in the Argonne Forest, 1916 / **72**
4.2. Trenches and bunkers, 1916 / **73**
4.3. Devastated landscape near Ypres, 1917 / **82**
4.4. Tree ravaged by artillery fire, 1915 / **83**
6.1. Malaria mortality in Italy, 1887–1950 / **115**
6.2. Agricultural development and malaria in Italy / **116**
6.3. Cartoon from ERLAAS pamphlet / **123**
8.1. Memorial at Gresse-en-Vercors / **152**
8.2. Cemetery at Saint-Nizier / **159**
8.3. Memorial at Col de la Chau / **161**
8.4. Ruins of Valchevrière / **163**

ACKNOWLEDGMENTS

Many people have helped to make the writing and publication of this book possible. From early on the German Historical Institute (GHI) in Washington, D.C., has provided invaluable support, sponsoring the conference in 2004 that inspired this volume and partially subsidizing its publication. The proposal for a conference exploring the connections between warfare and environmental degradation originated with former GHI director Christof Mauch, who also offered perceptive comments on early versions of several essays in this collection. GHI staffers Christa Brown and Bärbel Thomas made every aspect of the conference run like clockwork, while Dirk Schumann, David Lazar, Kelly McCullough, Christoph Strupp, Astrid Eckert, and Frank Zelko also contributed to its success. Throughout the editorial process, GHI editor Mary Tonkinson brought skill and commitment to both her work on the manuscript and her interactions with the contributors.

My colleagues at the University of North Florida have also been very generous with their time and expertise. Dale Clifford read portions of the text and shared her insights as a military historian, while Aaron Sheehan-Dean kindly provided a map from his own *Concise Historical Atlas of the U.S. Civil War* to illustrate Lisa Brady's chapter on Sherman's 1864–65 campaigns. Many thanks go to Oxford University Press for granting permission to reproduce the map from Aaron's book. In addition, David Wilson of North Florida's Center for Instruction and Research Technology not only created a map for a valuable chapter but also patiently tutored me in the use of GIS software.

I am grateful to the editor-in-chief at Texas A&M University Press, Mary Lenn Dixon, whose faith in this project has never wavered, and to her colleagues Linda Salitros, Diana Vance, and Jennifer Gardner Nader, who have done much to keep it moving forward and to help the book reach a broad readership. Sincere thanks are due to the anonymous readers who critiqued the original manuscript for the press and to Mark Cioc, Joseph Glatthaar, Martin Melosi, and Thomas Zeller for their prepublication endorsements. Finally, I want to thank my family and friends for their encouragement during the publication process. Without their unflagging support this book could not have been published.

WAR AND THE ENVIRONMENT

INTRODUCTION

Landscapes of Peace, Environments of War

Charles E. Closmann

MILITARY CONFLICT IS OFTEN a cause and consequence of environmental decline. In Darfur, for instance, climate change and desertification have exacerbated fighting between pastoralists and farmers, forcing more than two million people to flee their homes and villages in this arid region of Sudan. Huddled in camps along the Sudanese/Chadian border, the refugees are quickly exhausting scarce sources of water and timber, making a catastrophic situation even worse.[1] Elsewhere, the United States and countries of the former Soviet Union continue to learn about the vast quantities of chemicals, depleted uranium, and other residues of their respective military systems that litter battlefields, storage depots, and installations around the world. The cost to remediate these sites will reach into the billions of dollars, a legacy extending far into the twenty-first century. As these examples demonstrate, military operations (and occupations) can have devastating effects on natural resources, making a study of the relationship between war and the environment vitally important if societies are to avoid future conflicts and create a more ecologically sustainable world.[2]

The nine essays collected here examine the historical connections between war and the environment. In so doing, contributors ask several basic questions: How has warfare transformed the environment over time, where *environment* is defined as climate, landscape, flora, fauna, soil, water, and built settlements with which human communities interact? In what ways have environmental conditions changed the character of combat, including not only their influence on strategies and resource use but also on how humans experience and remember military conflicts? And, finally, how should the effects of warfare on ecosystems, cities, and other features of our physical surroundings be measured?[3]

For millennia scholars have been fascinated by the spectacle of war. Since at least the fifth century B.C., when Thucydides wrote his *History of the Peloponnesian War*, historians have chronicled the tactics of great generals and the ways in which war changes social institutions, topples monarchs, and expresses or enacts nationalist, racialist, and imperialist agendas. In fact, society has long recognized that to analyze war is to ponder the most radical and destructive of human actions and to confront a fundamental engine of world history.[4]

Yet until recently, few historians have examined the environmental consequences of military operations, despite war's enormous potential to transform nature. Even practitioners of the new military history have neglected this topic, despite their commitment to exploring the nexus between armies and society. Accordingly, works in this field consider such things as the role of women during war, concepts of masculinity among soldiers, and the plight of African Americans in the U.S. Navy, among many other topics.[5] But investigations into how conflicts over natural resources can lead to warfare or into the long-term effects of combat on woodlands, farms, and other features of the environment are rare.[6] Moreover, the few military historians who do consider the environment usually study factors such as terrain or weather—but only insofar as these elements constitute obstacles or advantages on the battlefield.[7]

Only in recent years have environmental historians been more willing to reflect on the connections between war and our physical surroundings. Emerging from the activist climate of the 1960s and '70s, the first scholars in the field of environmental history explored the perceived "ecological crisis" precipitated by capitalist expansion, mechanized agriculture, and industrial development. Consequently, the foundational texts of this discipline emphasized such issues as pollution, soil erosion, wilderness preservation, and Western definitions of nature.[8] Trends began to change in the 1980s and '90s as scholars and the media paid increasing attention to the ecological consequences of recent conflicts. Much of the research from this period also reflected an activist agenda, highlighting such widely publicized episodes as the deployment of Agent Orange in Southeast Asia during the 1960s or NATO's use of depleted uranium in Kosovo, but less polemical studies also appeared in the maturing field of environmental history.[9] Between 2004 and 2008, for example, annual conferences of the American Society for Environmental History (ASEH) featured panels on such topics as the environmental history of the American Revolution, militarized landscapes, and the mobilization of rivers during wartime.[10]

Two recent books exemplify the growth in this field. In *War and Nature: Fighting Humans and Insects with Chemicals from World War I to Silent Spring*, Edmund Russell argues that scholars have oversimplified portrayals of warfare and the control of nature as separate activities, asserting instead that chemical weapons and insecticides "coevolved" during the twentieth century at the intersection of military combat, technology, and industry.[11] The same technologies, delivery systems, and institutions—collectively known as the military-industrial complex—were used to destroy both insect and human enemies before, during, and after World War II. Moreover, reliance on chemical agents sometimes had unintended consequences. In Vietnam, for example, the U.S. military sprayed DDT to control malaria and win over the local population—

an effort that backfired when, in addition to killing mosquitoes, the pesticide killed many of the cats that helped to control crop-destroying rodents.[12]

A second notable book, *Natural Enemy, Natural Ally: Toward an Environmental History of War,* also highlights connections between warfare and society. Contributors to that collection explore topics ranging from the impact of combat on energy use in Finland to global forestry practices after World War I to the effect that fighting among the Mughal societies of India had on patterns of cultivation. Edited by Richard P. Tucker and Edmund Russell, this anthology shows that modern warfare often accelerated destruction of the earth's ecosystems, that any comprehensive assessment of war's impact on local and regional ecosystems can be made only from a historical perspective, and that military campaigns cannot always be separated from the agricultural or industrial practices of the societies engaged in conflict.[13]

The insight that warfare is so closely linked to practices such as farming and timber cutting has important implications for this volume. Among other things, the following collection represents an example of both the environmental history of war and the new military history. In addition, the historical connections between war, environmental change, and society suggest a more systematic way of understanding the environmental consequences of military operations. In *Seeing like a State: How Certain Schemes to Improve the Human Condition Have Failed,* James C. Scott considers the reasons why modern, state-sponsored projects have so often gone awry. In particular, Scott notes that ambitious programs to build new cities or harvest timber more efficiently have often been preceded by government attempts to simplify that which they seek to manipulate and control. Scott cites, for instance, the case of scientific forestry in Prussia, a program that increased yields of timber after the state catalogued and mapped the forests before introducing the widespread planting of Norway spruce. Ultimately, however, this project failed because the homogenous forests of nineteenth-century Prussia lacked the biodiversity that made woodlands resistant to disease; nor did the project take into account the "vast, complex, and negotiated social uses of the forest" that encouraged such diversity. As a result of ignoring complicated practical knowledge at the local level, the Prussian state suffered forest die-offs and lower timber yields in the long run.[14]

Scott's analysis of state power provides a valuable perspective on the relationship between war and the environment, since attempts by one society to control another will always have serious consequences for social relations and natural resources; and despite the best efforts of military planners or occupation authorities, those consequences are rarely predictable. No military organization has ever had such profound effects on the environment as the armed forces of the United States. As John McNeill and David Painter point out in a general survey of this topic (chapter 1), the army transformed the American continent in the nineteenth century by expanding the frontier, building infrastructure, and purchasing millions of resource-intensive weapons. In addition, the Corps of Engineers constructed harbors, canalized riverbeds, and opened the American West for trade and agriculture, while army troops contributed to the extinction of millions of bison. More recently, America's globally deployed military has required the construction of hundreds of bases worldwide, with massive ecological impacts on foreign countries. American forces have also used vast quantities of oil, metal, and other resources while promoting petroleum exploration and stimulating the growth of an enormous domestic chemical industry.

In chapter 2 Greg Bankoff also highlights the unforeseen effects of warfare on the natural world. Exploring links between colonial expansion, armed conflict, and Philippine forests, Bankoff observes that in tropical regions "the history of the forest is largely commensurate with the history of the societies that lived in and about it." Moreover, from precolonial times to the mid-twentieth century, the Philippines provided timber for Spanish, American, and Japanese forts, ships, weapons, and other military equipment. Bankoff draws on a rich variety of colonial letters, demographic data, and forestry reports to illuminate new aspects of this topic, arguing that "to talk only about the impact of war on the forests does an injustice to the complexity of that relationship." He notes, for example, that colonial foresters in the Philippines disproportionately harvested hardwoods such as teak, creating "genetic erosion" that not only devastated specific varieties of trees but also undermined woodland ecosystems in unforeseen ways.

In chapter 3, Lisa Brady studies the impact of the U.S. military on nature but from the perspective of an particular region and campaign. Writing on William T. Sherman's "march to the sea" in 1864, she describes how Union troops feasted on the abundant hogs, chickens, corn, and sweet potatoes of local farms while reveling in the dry weather, fine roads, and open countryside of rural Georgia. At the same time, however, Sherman's men intentionally devastated this landscape in order to destroy the agricultural foundations of the Confederacy. Sherman recognized that Confederate success depended on an "ecological" network of farms, railroads, and slaves; and he waged a brutal war against the land in order to disrupt these connections, demoralizing Confederate and Union troops alike. Brady helps us to recognize how combat against enemy soldiers and the environment shattered bodies, landscapes, and the human spirit.

This was especially true along the western front during World War I. In chapter 4, Dorothee Brantz draws on diaries, letters, and memoirs to describe the day-to-day experience of soldiers huddled in the muddy trenches of northern France. Brantz argues that the high-powered explosives, long-range artillery, and other "machine-age" weapons turned "landscapes of peace" into "environments of war." Moreover, the combined effects of powerful, technologically advanced weapons and the horrific conditions of trench warfare forced soldiers to engage with environmental hazards undreamed of by previous generations of soldiers. The first modern war, according to Brantz, World War I was also a conflict in which the environment was a "constitutive element," assaulting the bodies and senses of all combatants in ways that no military planners could have anticipated.

Moving from a discussion of landscapes and soldiers to organizations, Frank Uekötter examines the impact of wartime priorities on government conservation agencies in Germany. In chapter 5 he challenges recent scholarship that links Nazi Germany's ambitious conservation policies to the regime's racist ideology.[15] In Uekötter's view, German bureaucrats managed to impede projects likely to cause environmental damage not because these officials were motivated by ideological concerns but because "there were laws and regulations that had to be enforced, there were civil servants waiting for instructions, and there was an established routine carried over from the prewar years." In other words, it is too simplistic to say that warfare always shifts bureaucratic priorities away from environmental concerns. Uekötter demands that we consider not only the broad political goals of a modern regime but also the fact that individual bureaucracies are motivated by their own internal objectives, ideas, and dynamics.

In other cases modern conflicts have served as laboratories for scientific experimentation by governmental and nongovernmental institutions. In his essay on antimalaria campaigns during World War II (chapter 6), Marcus Hall shows how Italy's battle-scarred lowlands became a proving ground for new medicines, insecticides, and public-health strategies. As the war was ending, both U.S. and Italian malariologists tested new drugs on hospital patients and sprayed the newly discovered compound DDT over Sardinia's marshy lowlands, ignoring ethical considerations that might interfere with their research. Such experiments continued after the war was over. Funded by the Rockefeller Foundation and the United Nations, public-health experts in Italy established the Ente Regionale per la Lotta Anti-Anofelica di Sardegna, an agency known as ERLAAS, in order to wage an all-out war against malaria. ERLAAS spent millions of dollars and hired thousands of local farmers, who sprayed the entire island with DDT. According to Hall, the campaign succeeded in wiping out malaria while inadvertently also reducing the island population's natural immunity to disease and producing DDT-resistant mosquitoes. Long after malaria was eradicated, ERLAAS remained a testament to institutional inertia and the unintended effects of wartime public-health campaigns on Sardinia's people and its environment.

Even far from the battlefield, warfare has often accelerated new forms of natural-resource exploitation. In chapter 7, Robert Wilson traces the evolution of waterfowl policy in California during the 1930s, '40s, and '50s in order to show how the war effort drove the U.S. Fish and Wildlife Service (FWS) to integrate federal wildlife refuges into the state's agricultural system. Because total war demanded more acreage for rice planting in the Central Valley, a policy was developed to support this objective, with both negative and positive effects on migratory populations of ducks and geese. Blurring the lines between war and peace, the FWS used gunfire and harassment with aircraft to steer birds away from important crops and toward newly designated refuge land. In the long run FWS policies had uneven effects on waterfowl populations. During the 1950s, for example, agency experiments with insecticides killed numerous egrets, herons, and other wading birds while protecting rice crops and providing ample food for ducks and geese.

Shifting our focus to how societies remember warfare, Chris Pearson examines the ways in which French veterans of World War II modified the landscape to commemorate or to efface the experience of war.[16] Pearson studies this phenomenon in the Vercors redoubt of eastern France, where, in 1944, French partisans waged a short, bloody rebellion against German occupation troops. Taking a cultural-materialist approach, Pearson analyzes stone monuments and other artifacts to demonstrate how veterans "grounded" the memory of their resistance in the picturesque mountains, gorges, and ravines of the region, shaping the Vercors into an enduring symbol of resistance. From as early as 1945, however, private development has threatened the purportedly immutable nature of the Vercors, demonstrating in the process that most landscapes are themselves cultural artifacts.

In the volume's final essay Jeffry Diefendorf explores the effect of war on urban areas, a neglected topic in this field. He notes that postwar city planners saw the immense destruction of Berlin, Coventry, Tokyo, and other cities as an opportunity to rejuvenate urban areas according to modernist models. Informed by a widely shared conviction that big cities were "sick," soulless expanses of steel and concrete, these planners drafted plans to open up the urban core and expose residents to air, light, and greenery. In Berlin and Hamburg, for example, Allied air raids obliterated large portions of the urban landscape, creating a tabula rasa for planning schemes that predated the war by decades. The result in many cases was a mish-mash of modernist urban renewal, unplanned housing, and historical preservation, as idealistic designs clashed with property rights and other agendas. Diefendorf concludes with the observation that while all industrialized countries experimented with modernist planning after 1945, "where bombs fell, there was more change and more rapid change. For better or worse, bombed cities became the laboratories for postwar urban change."

Diefendorf's essay underscores some of the major themes that emerge in this volume. Warfare is an extreme expression of the power wielded by modern nation-states, which today possess unprecedented military resources and weaponry with nearly limitless destructive potential. Modern, machine-age battles have not only transformed landscapes and ecosystems but also the ways in which soldiers experience and remember combat. Warfare has often accelerated the pace of environmental change as natural resources became the target of artillery or ravaging armies, a means to sustain war-making capability, or the object of war itself. Finally, warfare sometimes "clears the table" for new forms of political economy or for ambitious, state-sponsored experiments in refashioning the natural and built worlds. All these facts make it vitally important for us to subject the environmental consequences of armed conflict—both deliberate and unforeseen—to rigorous investigation and analysis. If we are ever to realistically assess and effectively manage the impact of past, present, and future military actions on our natural resources and landscapes, then the relationship between war and the environment is one that cannot—and must not—be ignored.

Notes

1. Didrik Schanche, "Scarce Resources, Ethnic Strife Fuel Darfur Conflict," *NPR.org*, October 29, 2007, http://www.npr.org/templates/story/story.php?storyID=6425093, accessed June 17, 2008; "Sudan: Climate Change—Only One Cause among Many for Darfur Conflict," *Humanitarian News and Analysis, UN Office for Coordination of Humanitarian Affairs*, June 28, 2007, http://www.irinnews.org/Report.aspx?ReportID=72985, accessed June 17, 2008; Alfred de Montesquiou, "Experts: Darfur Faces Environment Crisis," June 22, 2007, http://www.washingtonpost.com/wp-dyn/content/article/2007/06/22/AR2007062201546.html, accessed June 17, 2008.

2. See J. R. McNeill, *Something New under the Sun: An Environmental History of the Twentieth-Century World* (New York: W. W. Norton, 2000), xxi–13, 357–62, and passim; Julian Perry Robinson, *The Effects of Weapons on Ecosystems* (New York: Published for the United Nations Environment Programme by Pergamon Press, 1979); and Seth Shulman, *The Threat at Home: Confronting the Toxic Legacy of the U.S. Military* (Boston: Beacon Press, 1992), 61–112.

3. For purposes of this collection, I have slightly modified the definition of *environment* and *environmental history* given in Carolyn Merchant's introduction to *Major Problems in American Environmental History*, ed. Carolyn Merchant (Lexington, MA: D. C. Heath, 1993), vii–x, 1–3.

4. The literature on war is enormous and far beyond the scope of a single footnote. Nevertheless, a few critical sources include Carl von Clausewitz, *On War*, ed. and trans. Michael Eliot Howard and Peter Paret (Princeton, N.J.: Princeton University Press, 1976); Lynn Montross, *War through the Ages* (New York and London: Harper and Brothers, 1944); Michael Carver, *War since 1945* (London: Weidenfeld and Nicolson, 1980); John Costello, *Love, Sex, and War: Changing Values, 1939–45* (London: Collins, 1985); Arthur Marwick, *The Home Front: The British and the Second World War* (London: Thames and Hudson, 1976); and Wilfried Loth, *The Division of the World, 1941–1955* (London: Routledge, 1988).

5. In a speech delivered to the fifty-eighth annual meeting of the American Military Institute in 1991, the eminent scholar Peter Paret defined the new military history as an "expansion of the subject of military history from the specifics of military organization and action to their widest implications, and also a broadening of the approaches to the subject, and of the methodologies employed" (quoted in John Whiteclay Chambers, "The New Military History: Myth and Reality," *Journal of Military History* 55 [July 1991]: 395–406). For a more recent survey of the new military history, see Robert M. Citino, "Military Histories Old and New: A Reintroduction," *American Historical Review* 112 (October 2007): 1070–90. Works that exemplify the new military history include Elizabeth D. Leonard, *Yankee Women: Gender Battles in the Civil War* (New York: W. W. Norton, 1994); Donald R. Shaffer, *After the*

Glory: The Struggles of Black Civil War Veterans (Lawrence: University Press of Kansas, 2004); and Steven J. Ramold, *Slaves, Sailors, and Citizens: African Americans in the Union Navy* (DeKalb: Northern Illinois University Press, 2002).

6. This conclusion is based in part on my examination of the tables of contents for all issues of the *Journal of Military History* from 2005 to 2008. During this review I found no articles with an explicitly environmental emphasis, although I identified at least one with a focus on military geography; see Lorraine White, "Strategic Geography and the Spanish Habsburg Monarchy's Failure to Recover Portugal, 1640–1668," *Journal of Military History* 71 (April 2007): 411–40. Moreover, Robert Citino makes no mention of environmental issues in his valuable essay cited above.

7. Richard P. Tucker and Edmund Russell, eds., *Natural Enemy, Natural Ally: Toward an Environmental History of Warfare* (Corvallis: Oregon State University Press, 2004), 1–13.

8. Environmental history now has a relatively well established historiography. In addition to the works mentioned in this essay, the antecedents of this field lie in the disciplines of geography, which considers the influence of physical environments on human social structures; the French Annales School, which explores structural trends shaping human history; scientific ecology, which emphasizes relationships between organisms and their surroundings; and the disciplines of archaeology and anthropology. For a brief summary, see K. J. W. Oosthoek, "What Is Environmental History" (posted in 2005) in the "Essays" section of the *Environmental History Resources* Web site, http://www.eh-resources.org/environmental_history.html, accessed June 17, 2008; and Douglas R. Weiner, "A Death-Defying Attempt to Articulate a Coherent Definition of Environmental History," *Environmental History* 10 (July 2005): 404–21. For additional perspectives on the evolution of environmental history, see the editors' introduction to Tucker and Russell, eds., *Natural Enemy, Natural Ally*, 1–13. The first well-known works in this field illustrated this emphasis on domestic affairs. See, for example, Roderick Nash, *Wilderness and the American Mind* (New Haven, Conn.: Yale University Press, 1967); Donald Worster, *Dust Bowl: The Southern Plains in the 1930s* (New York: Oxford University Press, 1979); and Donald Worster, *Nature's Economy: A History of Ecological Ideas* (Cambridge: Cambridge University Press, 1977). For a seminal essay in this field, and one that challenged society to rethink Western attitudes toward the natural world, see Lynn White Jr., "Historical Roots of Our Ecological Crisis," *Science* 155 (1967): 1203–07.

9. Rauno Lahtinen and Timo Vuorisalo, "'It's War and Everyone Can Do as They Please!' An Environmental History of a Finnish City in Wartime," *Environmental History* 9 (October 2004): 679–701. My search of the literature on war and the environment reveals numerous works published since the mid-1980s on the potential environmental consequences of nuclear war, the impact of defoliant use during the Vietnam War, the consequences of the Gulf War, and the use of depleted uranium in the Balkans during the 1990s. For just a few of the recent works that address these issues, see Lester R. Brown, ed., *State of the World, 1991: A Worldwatch Institute Report on Progress toward a Sustainable Society* (New York: W. W. Norton, 1991); Sheryl J. Brown and Kimber M. Schraub, eds., *Resolving Third World Conflict: Challenges for a New Era* (Washington, DC: U.S. Institute of Peace Press, 1992); Constantine S. Inati, *Iraq: Its History, People, and Politics* (Amherst, NY: Humanity Books, 2003); United Nations Environment Programme, United Nations Centre for Human Settlements, *The Kosovo Conflict: Consequences for the Environment and Human Settlement* ([Geneva]: UNEP-UNCHS, 1999).

Introduction

10. Programs for the annual conferences of the American Society of Environmental History from 2003 onward are available online at http://www.aseh.net/conferences/conference-archives.

11. Edmund Russell, *War and Nature: Fighting Humans and Insects with Chemicals from World War I to Silent Spring* (New York: Cambridge University Press, 2001), 226–36.

12. Ibid., 226; McNeill, *Something New under the Sun,* 203, 260, 339.

13. See the following essays in Tucker and Russell, eds., *Natural Enemy, Natural Ally*: Richard P. Tucker, "The World Wars and the Globalization of Timber Cutting," 110–41; Stewart Gordon, "War, the Military, and the Environment: Central India, 1560–1820," 42–64; Mark Fiege, "Gettysburg and the Organic Nature of the American Civil War," 93–109; Simo Laakkonen, "War—An Ecological Alternative to Peace? Indirect Impacts of World War II on the Finnish Environment," 175–94; William M. Tsutsui, "Landscapes in the Dark Valley: Toward an Environmental History of Wartime Japan," 195–216.

14. James C. Scott, *Seeing like a State: How Certain Schemes to Improve the Human Condition Have Failed* (New Haven, Conn.: Yale University Press, 1998), 1–15, esp. 13.

15. Franz-Josef Brüggemeier, Mark Cioc, and Thomas Zeller, eds., *How Green Were the Nazis? Nature, Environment, and Nation in The Third Reich* (Athens: Ohio University Press, 2005).

16. For an excellent introduction to this topic, see George L. Mosse, *Fallen Soldiers: Reshaping the Memory of the World Wars* (New York: Oxford University Press, 1990).

CHAPTER 1

THE GLOBAL ENVIRONMENTAL FOOTPRINT OF THE U.S. MILITARY, 1789–2003

J. R. McNeill and David S. Painter

Since the origins of states some five thousand years ago, military organizations, usually armies but occasionally navies as well, have played central roles in human affairs. The chief function of most states in times past was making war. By the seventeenth century, many, including the Mughal Empire, the Ottoman Empire, or the France of Louis XIV, spent half or three-quarters of their revenues on their military machines. When large states found themselves, or put themselves, at the center of international political struggles, they invariably invested heavily in military capacity, as the United States has done since the 1890s. In such situations military investment became a major factor in the national or imperial economy. The military's political power and social importance affected decisions and outcomes in every realm of society, from education to infrastructure. Given the centrality of military organizations, especially in imperial states, it is clear that no analysis of social, economic, or political trends is complete without due consideration of the military. The same, we contend, is true of ecological trends.

This chapter seeks to assess the ecological influence of the American military in general terms. It begins with the creation of the American republic but concentrates on the period since 1890, the era in which the United States became a great power in the international system and eventually the sole superpower. The U.S. military directly affected the environment at home and abroad on many levels. It spearheaded the campaign to eliminate Amerindian power in the national territory, preparing it for settlement by Euro-Americans and, to a lesser extent, by African Americans. It built new infrastructure. Its demand for reliable and increasingly potent weaponry spurred the development of an efficient armaments industry and eventually led to the creation of a sprawling military-industrial complex. It acquired an archipelago of military bases around the world. It pioneered the development of nuclear weapons and nuclear power. In countless ways, large and small, the U.S. military has affected the environment both in the United States itself and around the world.

Ecological impact is not of course unique to the American military.[1] Every military organization has had such effects, and those of great imperial states in centuries past had major consequences. Certainly the Soviet military after 1918 exercised a powerful ecological influence within its own territory, within those of its satellites in Eastern Europe, and on its bases overseas. Indeed the Soviet military was particularly careless with respect to, for example, nuclear weapons.[2] But no power has had ecological effects quite so pervasive as those of the United States.

The Size and Mission of the U.S. Military since 1789

The American republic was born of war. Between 1775 and 1781 a few thousand armed men managed, with help from France, to resist the power of Great Britain and create a new nation. Instantly thereafter the American military shrank to near oblivion: in 1789 it had a grand total of 718 men in uniform. Over the next century the main job of the U.S. Army—its navy remained tiny—was to expand the national territory westward, at the expense of Amerindians, of Spain, and, after 1845, of Mexico. Its second responsibility was to safeguard that territory from possible British attack, which became an immediate concern during the War of 1812, when the two countries fought again. Britain retained an interest in North America and colonial possessions in Canada, so that at times frictions developed over land in Oregon, California, and Texas. The third function of the American military was to hold the republic together, which became its chief task during the Civil War of 1861–65. None of these duties, except fighting the Civil War, required large investments of men or money. The Amerindians and Mexicans were not formidable foes, and Britain was far away and always engaged with events more urgent or closer to hand than those in North America. So the American military remained small until 1898, except for the Civil War years.

Table 1.1 lists the total number of U.S. service personnel at five-year intervals and also includes data for three years that witnessed unusual but fleeting expansions: 1848, 1898, and 1918. It shows the long-term evolution of manpower in all branches of the military (army, navy, marines, air force), revealing the 1890s and 1940s as pivotal moments of sustained growth when the American military took on new missions. The Civil War (about a forty- or fiftyfold expansion from immediately preceding peacetime levels) was followed by decades of retrenchment, despite the numerous "Indian wars" of the 1870s and 1880s. However, the more modest (sixfold) expansion that came with the Spanish-American War of 1898 resulted in long-term growth in the manpower of the military. After 1898 the number of men in uniform never fell below 100,000. After the great mobilization of World War I (roughly a twentyfold expansion) it never fell below 250,000. And after the greatest of all mobilizations, that for

TABLE 1.1 MANPOWER OF U.S. MILITARY ON ACTIVE DUTY, 1789–2000 (IN THOUSANDS)

Year	Personnel	Year	Personnel
1789	0.72	1900	126
1795	5.3	1905	108
1801	7.1	1910	139
1805	6.5	1915	174
1810	11.5	1918	2,897
1815	40.9	1920	343
1820	15.1	1925	252
1825	11.1	1930	256
1830	11.9	1935	252
1835	14.3	1940	458
1840	21.6	1945	12,126
1845	20.7	1950	1,460
1848	60.3	1955	2,935
1850	20.8	1960	2,476
1855	26.4	1965	2,655
1860	27.9	1970	3,066
1865	1,062.8	1975	2,128
1870	50.3	1980	2,051
1875	38.1	1985	2,151
1880	37.9	1990	2,044
1885	39.1	1995	1,518
1890	38.7	2000	1,384
1895	42.2		
1898	236.0		

Source: U.S. Department of Commerce, *Statistical Abstract of the United States, 2001* (Washington, D.C.: U.S. Department of Commerce, 2001); U.S. Department of Commerce, *Historical Statistics of the United States: Colonial Times to 1970* (Washington, D.C.: U.S. Department of Commerce, 1971).

World War II (a fiftyfold expansion), it never again fell below a million and usually stood well above two million until the 1990s. These figures reflect the new positions in world politics taken by the United States in the 1890s and the 1940s. Since the 1970s the military has been gradually shifting toward a more technology-based and less labor-intensive structure; its declining manpower since 1992 reflects the end of the Cold War and the transition to a new set of challenges.

The Environmental Impact of the U.S. Military to 1890

Before 1890 the U.S. military had little impact outside the North American continent. In North America its efforts brought widespread ecological change in three main ways: through frontier expansion; through infrastructure construction; and, rather indirectly, through purchases of weaponry. Between 1790 and 1890 the U.S. Army fought countless "Indian wars." Their stated purpose was to safeguard U.S. citizens and to subdue uncooperative Amerindians, opening the way for settlement by farmers and, in the South until 1861, by planters and their slaves as well. The civilian population of the United States might have accomplished this with minimal assistance from the army, as the Canadian experience suggests. But the army sped up the process and made it less costly for civilians. By 1890, when the U.S. Army crushed the last remnants of organized resistance among the Plains Indians, Amerindian military power was decisively and definitively broken.

The Army also participated in the federal registration of lands appropriated from Amerindians as well as in the sale and distribution of properties to white settlers in these regions. Army engineers conducted surveys for the canals and railroads that crisscrossed the United States in the nineteenth century, the most important factor in promoting settlement. It thus accelerated an ecological transformation of the territory west of the Appalachian Mountains—that is, over more than five-sixths of the country—because a new system of human ecology came with the settlers. Cotton and wheat spread westward as far as rainfall agriculture could go, replacing the various combinations of horticulture, hunting, and gathering that had previously sustained Amerindian peoples. The army also played a significant role in reducing bison populations from perhaps twenty or thirty million to a thousand or fewer, opening up the semiarid heartland of the United States to cattle ranching after the 1870s. Destruction of the bison herds also eliminated "the Indians' commissary," as one army general put it.[3]

All this amounted to a great ecological shift, driven mainly by changes in land use but also by increased population density of the sort common on nineteenth-century settlement frontiers. Alfred Crosby describes this process in detail, calling it "ecological imperialism" and applying it to the European colonization of Australia, New Zealand, Argentina, southernmost Brazil, as well as to North America.[4] But it is well to recognize that in all of these cases the process of demographic replacement and the establishment of a new system of human ecology were accomplished not only by microbes and alien diseases but also by military force. Nowhere was this truer than in the United States (although the cases of New Zealand and Argentina came close).

The second means by which the U.S. military affected North American ecology was through its commitment to infrastructure development. The principal agent here was the U.S. Army Corps of Engineers, founded in 1802. In addition to surveying work, the corps constructed much of the country's navigational infrastructure. It built and dredged harbors on the seacoasts and on the Great Lakes. It dug canals, created locks, and channelized riverbeds, helping to make the Ohio and Mississippi rivers great arteries of commerce by 1850 and the Missouri River a significant commercial waterway by 1880. It also built roads, lighthouses, piers, docks, bridges, aqueducts—even the U.S. Capitol building. These undertakings altered local landscapes and ecologies fundamentally, notably in the case of river floodplains. But more importantly, they greased the wheels of trade and accelerated the pace of settlement in the age before railroads. At the time of the Civil War, for example, most U.S. cotton passed through New Orleans along waterways improved for navigation by the corps.[5] Even in the railway age, river and canal transport remained the cheapest way to move bulk goods; so the corps' projects, by turning rivers into navigable waterways, helped to turn the nation's forests into timber and its prairies into pasture and croplands.

The third main way in which the U.S. military shaped the North American environment before 1890 was through its role as a purchaser of armaments. The military influence on American—and world—industrial history was large even though the army and navy were small customers except during the Civil War. The U.S. Army was the first buyer of weapons composed of interchangeable parts, an idea that originated in France but encountered widespread resistance from artisans. The army overcame that obstacle by sponsoring experiments in precision manufacturing at an armory in Harpers Ferry, Virginia. By 1850 American metalworkers could cut and mold parts to very fine tolerances and then use unskilled labor to assemble these components into reliable weapons. Americans called this the "armory system," while elsewhere in the world it became known as the "American system" of manufacture.[6]

The American system first took hold in the field of metallurgy and in the engineering and construction of machinery, branches of industry in which America would soon lead the world. But its principles were rapidly applied to the operations of a broad range of industries, including even the great stockyards of Chicago, where cattle were systematically "disassembled." It led eventually to the electrified assembly line of Henry Ford. Presumably this system would have become the standard in manufacturing even without the impetus provided by the U.S. Army, but its establishment as accepted practice was undoubtedly hastened by the military's involvement. Because this type of design made more sense in the manufacture of weapons, which were vulnerable to damage in combat and thus often required spare parts, the military had a strong motive to develop the American system and provided the crucial demand that kept nineteenth-century armories in operation.

To some extent, then, the Army's ordnance orders were responsible for the technical and management breakthroughs that underwrote American success in metallurgical industries, with all the environmental effects that implies. Those ranged from the landscape disruptions inevitable in the mining of iron ore and coal to the prodigious air and water pollution generated by the nineteenth-century iron and steel industry.

From the 1890s to World War II

The 1890s were a pivotal decade for U.S. foreign policy, and the changes that occurred then had major impacts on the size and mission of the U.S. military. Although many historians once viewed the Spanish-American (or, more accurately, the Spanish-American-Cuban-Filipino) War and the U.S. acquisition of an external empire as an aberration, most historians now view expansion, not isolation, as the dominant characteristic of nineteenth-century U.S. foreign policy. From the American Revolution to the beginning of the twentieth century, American leaders believed that territorial and economic expansion were indispensable to the survival and success of the nation's political and economic systems. Protected by its geographical distance from the great European powers, by the balance of power in Europe, and by having neighbors who posed no threat to it, the United States had no need for a large military establishment. Nor did it have any need for a large navy to protect and advance its commerce because British policy and Britain's Royal Navy willingly bore the burden of building and maintaining an open world economy.

By the 1890s, with expansion across the continent complete, a rapidly growing population, and a rising industrial economy in search of markets and raw materials, U.S. leaders began to look abroad for "new frontiers" to conquer in order to maintain stability and prosperity. Technological changes—in particular the shift from wind-driven, wooden-hulled ships to steam-powered, steel-hulled ships and the advent of the "new imperialism," whereby European colonial expansion legitimized U.S. (and Japanese) efforts—provided additional impetus for changes in U.S. foreign and military policy. Drawing on the writings of naval theorist Alfred Thayer Mahan, the United States began to develop a modern navy capable of projecting American power across the seas. The U.S. naval buildup, which began in the 1880s, not only propelled the United States into the top ranks of naval powers but also forged close ties between the U.S. military and the industries, in particular the steel industry, that profited from its expansion. By 1893 most analysts ranked the U.S. Navy seventh in the world; by 1914, third; and by the end of World War I, equal to Britain's Royal Navy.

At the time of its buildup, the U.S. Navy, like those of other naval powers, depended on coal as its fuel. In the years 1892–93 the Navy used only about seventy thousand tons of coal, but by 1919 it needed nearly four million tons, and had become one of the main customers for the Appalachian coal mines, although the Navy never accounted for more than 1 percent of national coal consumption. Indirectly, of course, the Navy was responsible for further coal production because its orders spurred shipbuilding, which spurred steel production. In any case, the steam-powered navy added slightly to the environmental consequences of mining and burning coal. The most serious effects (in the age of the steam-powered navy) included the pollution of streambeds, the proliferation of slag heaps, and increased air pollution—all local in scope if sometimes acute in nature.[7]

Another important consequence of the shift from sail to steam was the need for coaling stations in distant locales. To protect coaling stations, bases were necessary, and the need for bases led directly to territorial acquisitions. Thus, although the essential character of U.S. expansion changed from territorial to commercial, a distinct element of territorial expansion remained. The United States acquired Midway Island in 1867 and in 1884 was granted exclusive rights to build a naval base at Pearl Harbor in what was then the Kingdom of Hawai'i. American planters seized control of the islands in 1893, and five years later, during the war with Spain, the United States annexed Hawai'i. After dredging the mouth of the Pearl River to permit large ships to enter the harbor, the United States developed a coaling station and a major naval base at Pearl Harbor, and following World War I, built airfields there.[8]

The desire for a base in East Asia to protect U.S. access to China was a key motivation behind the annexation of the Philippines following the war with Spain. The United States quickly established a naval base at Subic Bay, northwest of Manila. And, as in Hawai'i, the United States also constructed a large airfield north of Manila in 1919. The shift of the U.S. Navy from coal to oil, by giving warships greater range, allowed the U.S. Navy to "cover" East Asia from Hawai'i. This development turned the Philippines from a strategic asset to a liability, and in the Five Power Naval Treaty signed at the Washington Conference in 1922, the United States agreed not to build any new naval or military installations in its territories west of Hawai'i. The United States also acquired Guam from Spain but did not construct any major bases on the island before World War II.

American acquisitions of Spanish territory took place close to home as well as across the Pacific. Although the United States insisted on Cuban independence, it occupied Cuba after the war with Spain. Among the conditions it set for ending its occupation was that Cuba cede the United States lands necessary for coaling or naval stations. At the end of 1903 the United States took control of about 130 square kilometers of land and water at Guantánamo Bay on the southeastern tip of Cuba. Despite Guantánamo's fine deep-water port and location near the Windward Passage, the shortest sea route from the east coast of North America and Europe to the Caribbean, the United States did not develop a sizable military base there until World War II. Similarly, though the United States gained the right to build bases in Puerto Rico, which it also took from Spain in 1898, it did not establish any big bases there until World War II.

By the 1890s American strategists saw control of Hawai'i and bases in the Caribbean as necessary to protect the approaches to a canal across the Central

American isthmus connecting the Atlantic and Pacific oceans. On November 18, 1903, just two weeks after Panama achieved independence from Colombia with assistance from the United States, the Hay–Bunau–Varilla Treaty was signed in Washington, D.C. This agreement authorized construction of an interoceanic canal and granted control in perpetuity over a five-mile zone of land on each side of the canal to the United States, as "if it were the sovereign of the territory." The United States quickly exercised its right by building a series of military fortifications and bases in the canal zone. These acquisitions laid the foundations for the post–World War II U.S. system of military bases.

The construction of ports, fortifications, airfields, barracks, maintenance and repair shops, and fuel-storage facilities had significant local consequences, but the relatively small size of the bases meant that on the one hand their overall environmental impact remained minor. On the other hand, U.S. involvement in Cuba and Panama led the United States and its military to take a greater interest in tropical medicine and disease control. American and Cuban doctors unraveled the mysteries of yellow fever between 1890 and 1920, an epidemiological advance resulting in successful public-health measures—specifically mosquito control—that dramatically changed the disease environment of the American tropics. Developing the means to identify and kill the mosquitoes that spread malaria and yellow fever proved crucial to the construction of the Panama Canal. The U.S. military also undertook extensive road-building programs as well as sanitation and disease-control efforts in Cuba and other Caribbean nations temporarily occupied by American forces: Nicaragua, 1912–25 and 1926–33; Haiti, 1915–34; and the Dominican Republic, 1916–24. These U.S. military activities had significant environmental consequences, first in and around the military installations themselves and eventually throughout the Caribbean and Central America. In particular, the military presence in the Caribbean and Central America, as well as in the Pacific, facilitated U.S. investment in tropical agriculture (sugar and banana plantations, for example), with the attendant ecological reshuffling that such investment entailed.[9]

The U.S. Navy expanded greatly during World War I, in part because President Woodrow Wilson wanted to have a strong position from which to mediate an end to the conflict. As he told his confidant, Col. Edward House, in September 1916, "Let us build a navy bigger than [Great Britain's] and do what we please."[10] Just before the war, the United States had begun to switch its navy from coal to oil fuel. Oil was more efficient, provided more power, and was easier to transport and store. Oil was also essential to the major military innovations of World War I—the submarine, airplane, tank, and motorized transport. By increasing the mobility of military forces, these innovations enlarged the scale and the scope of their destructive power.

The "motorization" of warfare continued throughout the interwar period, and by World War II all the major weapons systems were oil powered. The United States succeeded in checking a nascent naval arms race following World War I with naval limitations treaties in 1922 and 1930, but arms-control efforts broke down in the 1930s, forcing the United States to resume buildup of its naval forces. To keep abreast of rapidly developing airplane technology at a low cost, the U.S. military adopted a policy of funding the development and testing of prototypes rather than sinking funds into the acquisition of aircraft that would soon be obsolete. Although the U.S. Army remained small by great power standards, it maintained and developed the officer corps that led the vastly enlarged army of World War II. Finally, U.S. military potential, based on its huge and technologically advanced industrial economy and abundant resources, far exceeded that of any potential rival. In particular, the U.S. steel, automobile, airplane, and oil industries provided an unrivaled basis for wartime mobilization.

The motorization of the U.S. military did not have a large immediate impact on the environment because, with the exception of the navy, America's armed forces remained relatively small until 1941. Nevertheless, the growing reliance of the U.S. military machine on petroleum, combined with burgeoning civilian oil consumption, sharpened the military's interest in overseas sources of oil and encouraged involvement of U.S. oil companies in Latin America, the Middle East, and the Netherlands East Indies. It also increased American society's dependence on oil, and all the ecological implications thereof, from oil drilling and transport, with their infrastructure and their spills, to oil combustion and its pollution effects.

While the military merely helped to accelerate the transition to oil, there were other things it and only it could do: the mass production of chemical weapons, for example. These would unleash a new range of environmental changes and hazards, both directly, as weapons, and indirectly, in their impact on the pesticide industry. Chemical weapons, in the form of noxious gases, have a long history but mattered very little until World War I. At Ypres, in 1915, the German Army first used poison gas on a large scale, killing about five thousand Allied soldiers. The era of modern chemical warfare had begun. The Americans lagged behind, partly because the military resisted the idea of using gas (as most militaries did for a while) and partly because the American chemical industry before World War I was small and unsophisticated. Both circumstances changed with the war.

The difficulties of wartime trade with Germany, whose chemical industry was then by far the largest and most technologically advanced in the world, paradoxically proved a boon to American chemical producers. They expanded operations rapidly to fill the void created by lost imports. As the United States prepared for war, its army slowly acceded to the salesmanship of chemical entrepreneurs and began to produce chlorine and mustard gas. By the end of the war the United States had built a chemical-weapons industry with six times the capacity of the Germans."[11]

The expertise developed with lethal chemicals during World War I transferred easily to warfare waged against insect pests. The chemical industry marketed new pesticides to housewives and farmers. Agricultural pesticides had existed in the nineteenth century on a small scale, but now they became cheaper and more effective. Furthermore, another technology much advanced by World War I, the airplane, combined with pesticides very effectively; in the 1920s Americans pioneered aerial spraying with lethal chemicals, a practice known as crop dusting. This inaugurated a new era in American croplands and their waterways, one in which chemicals—and evolved resistance to them—became a determining factor in the evolutionary success of plants and animals generally and of insects in particular.

Between the wars chemical-weapons development languished in the United States. The Geneva Protocol of 1925 outlawed the use of chemical and biological weapons. When the United States entered World War II, American generals considered using poison gas but decided against it, partly because President Roosevelt objected. Only the Japanese in China and the Germans in their concentration camps used poison gas during World War II (although Churchill was at times sorely tempted).[12]

But the American war effort produced two notable chemical weapons, one aimed against human enemies and one against insects: incendiary bombs and DDT. Incendiary bombs had been used in colonial wars in the 1920s,[13] but the Americans in World War II refined the chemical formulas and undertook mass production. They had no incendiary bombs when Pearl Harbor was attacked in December 1941; by December 1945 they had stockpiled a quarter billion. In the latter stages of the war enormous numbers of these bombs rained down upon targets in Germany and Japan, burning Hamburg, Dresden, and Tokyo comprehensively. Incendiary bombs killed far more Germans and Japanese, and flattened much more urban landscape, than did the two atomic bombs dropped over Japan.

In the Pacific theater the Americans encountered a foe more dangerous than the Japanese: malaria, which resulted in eight times as many American casualties as were produced by combat. The U.S. military health authorities experimented with various remedies before finding the ultimate weapon in 1943. A Swiss chemist had discovered the insecticidal properties of DDT in 1939, and his employer offered the formula to both the American and the German armies. The Germans deployed DDT on a modest scale to control typhus, a louseborne disease. The Americans used DDT on a huge scale against typhus and malaria; the chemical was wondrously effective against the malarial mosquito, and its application saved hundreds of thousands of lives during the war and many millions more in the decades after it. In the United States and around the world, DDT was widely and successfully used after World War II, especially in mosquito control.[14]

Eventually, however, DDT produced resistant mosquitoes and an "arms race" between chemists and mosquito evolution, one that continues to be waged. DDT, as a persistent organic compound (meaning one that does not degrade easily), also lingers in ecosystems. It accumulates in the tissues of birds, fish, livestock, and humans, causing die-offs and health problems. The

United States banned its use in 1972, but its presence in the environment and in animal tissues meant that it remained a hazard, especially to predatory fish, into the 1990s. Still employed around the world, DDT has been found even in the tissues of Antarctic penguins, thousands of kilometers from where it was sprayed.[15]

Military Infrastructure and the Environment since 1941

Both the size and power of the U.S. military escalated tremendously during World War II. Although the level of the U.S. armed forces dropped sharply from 12.1 million in 1945 to 1.7 million by mid-1947, the United States still possessed the world's most powerful military machine in the ensuing years. The U.S. Navy, bolstered by wartime shipbuilding, controlled the seas; U.S. air power dominated the skies; and the United States alone possessed atomic weapons and the means to deliver them. Following the Soviet Union's acquisition of atomic weapons in 1949, the United States sought to maintain strategic superiority through increased production of atomic weapons and development of the hydrogen bomb (see below). The Korean War (1950–53) led to another massive increase in U.S. military spending, and the United States expanded its armed forces by over a million troops and sharply increased production of aircraft, ships, combat vehicles, and other conventional weapons. The U.S. nuclear arsenal also grew dramatically from roughly one thousand warheads in 1953 to approximately eighteen thousand by 1960. Although the United States began to reduce the size of its conventional forces and slow the rate of production of nuclear warheads in the 1970s, each new generation of weapons possessed more destructive power, consumed more energy, and had a greater environmental impact.

The experience of World War II changed the way the American public thought about national security interests and requirements. Drawing on what they believed to be the lessons of the 1930s, U.S. leaders sought to create and maintain a favorable balance of power in Europe and Asia, to fashion an international economic order open to U.S. trade and investment, and to maintain the integration of the third world in the world economy in an era of decolonization and national liberation. To achieve these goals, U.S. leaders considered a global network of military bases essential to national defense. These bases would also serve to deter aggression against American interests and to influence events abroad by projecting U.S. power into potential trouble spots. Maintaining nuclear superiority was also seen as a crucial element of national defense strategies, which included both preempting attacks on the United States and preventing Soviet advances elsewhere in the world. These policies were designed not only to ensure the physical security of the United States and its allies but also to preserve a broadly defined "American way of life" by constructing an international order that would be open to and compatible with U.S. interests and ideals.[16] Until the deployment of B-52 intercontinental bombers in the late 1950s and long-range land- and submarine-launched ballistic missiles in the 1960s, the U.S. nuclear deterrent was heavily dependent on bombers and ballistic missiles that could not reach the Soviet Union from the United States. Likewise, conventional forces needed to be close to potential theaters of conflict if they were to be able to influence events. These geographical and technological factors made the United States dependent on bases in the territories of its allies.

The United States developed an archipelago of military facilities around the world covering around 810,000 hectares or something over two million acres, making it nearly the size of Lebanon or Connecticut. By the mid-1960s the United States had approximately 375 major military bases in foreign countries and 3,000 minor military facilities spread around the world. The largest single concentration of U.S. troops was in Europe, and over two-thirds of these were in West Germany. There were also large concentrations of U.S. forces in Japan (including Okinawa), the Philippines, South Korea, and Great Britain. By the end of the Cold War there were more than 525,000 U.S. troops permanently based overseas. A decade later the number of U.S. troops deployed overseas had fallen to around 235,000, with 109,000 in Europe, 93,000 in Asia, and 23,000 in the Persian Gulf. In addition, the United States had increasingly begun to substitute the pre-positioning of equipment, munitions, and fuel for the overseas deployment of military manpower. This new policy became standard procedure in the Middle East, where the presence of large numbers of U.S. military personnel was a particularly sensitive issue.[17]

The size and nature of American forces abroad dramatically affected the ecosystems in and around overseas military bases, which were usually exempt from both local and U.S. environmental regulations. In addition to such special classes of lethal by-products as radioactive material, high explosives, chemical weapons, and rocket fuels, U.S. military bases generated massive amounts of hazardous wastes as part of their normal operations. The routine maintenance of the vast numbers of ships, aircraft, combat and support vehicles, and weaponry produced such pollutants as used oil and solvents, polychlorinated biphenyls (PCBs), battery and other acids, paint sludge, heavy metals, asbestos, cyanide, and plating residues. Sometimes the size of small cities, U.S. bases also produced large amounts of ordinary garbage, medical wastes, photographic chemicals, and sewage.[18] Often these wastes, toxic and nontoxic alike, were disposed of casually. In a 1991 study of seven U.S. bases in Europe and the Pacific, General Accounting Office inspectors discovered ground or water contamination at five of the seven bases and four of the six central waste-disposal facilities examined.[19]

Training and maneuvers altered local landscapes, consumed vast amounts of energy, and contributed to air pollution, including the carbon dioxide buildup in the atmosphere that is responsible for climate change. The amount of land required for military maneuvers has increased dramatically since World War II as the size, speed, and complexity of military weapons and equipment have grown. In addition to the obvious impact of gunnery and bombing exercises, which cause not only immediate damage but also long-term hazards due to unexploded ordnance, the effects of large numbers of heavy vehicles on the land could be devastating to soil, vegetation, and waterways.[20]

Some of the worst environmental damage occurred at U.S. bases in the Philippines and Panama. The two largest bases in the Philippines, Clark Air Base and Subic Bay Naval Base, covered more than 77,000 hectares (or upward of 190,270 acres) and served as rear areas during the Korean War and major staging areas for the U.S. war in Vietnam. They continued to play an important role in U.S. strategy in the 1970s and '80s as a link to the Indian Ocean and Persian Gulf and as a means of countering the expanded Soviet naval presence in Asia. Following extensive damage to Clark and Subic Bay by the eruption of Mt. Pinatubo in 1991, the Philippine Senate refused to renew its agreement with the United States, leading to closure of these bases. More than four decades' worth of operations at the Clark and Subic Bay bases had generated huge amounts of hazardous waste, much of which was dumped into open drains or buried in unprotected landfills. Moreover, leakage from underground petroleum tanks and from the petroleum pipeline connecting Subic Bay to Clark Air Base had contaminated local groundwater, and unexploded ordnance continued to claim Filipino victims long after the U.S. military's departure.[21]

During World War II the U.S. base system in Panama developed into one of the world's greatest concentrations of American military force overseas, peaking at 63,000 troops in 1943. While the Canal facilitated the deployment of U.S. naval forces during World War II, postwar aircraft carriers were too large to pass through the locks, and the role of U.S. forces in Panama shifted from enhancing the nation's global reach to maintaining U.S. influence in Latin America. By the 1970s there were eleven major U.S. military installations in Panama. The 1977 Panama Canal treaties not only returned control of the canal to Panama by the end of the century but also required the withdrawal of the U.S. military presence. In July 1999 the Southern Command moved to Miami and Puerto Rico, and in December of that year the United States turned over its last military bases in Panama.

During World War II the U.S. military conducted numerous tests of chemical weapons in Panama, especially on San José Island in Panama Bay, which acquired the nickname "Test-Tube Island." After the war, the military continued to store and experiment with chemical weapons in Panama, including environmental tests of sarin and VX nerve gas, which sought to determine the effects of storage in tropical conditions on these deadly agents. The military also tested the dangerous defoliant Agent Orange in Panama and stored and tested depleted uranium munitions.[22] In addition to storing, testing, and disposing of chemical weapons in several areas of the country, the U.S. military maintained several firing ranges in Panama. According to one estimate, "the United States left more than one hundred thousand pieces of unexploded ordnance in Panama."[23]

Despite the scale of the overseas bases, the environmental damage caused by the U.S. military was greater at home than abroad. Less than a fifth of U.S. military manpower was stationed overseas, so the main impact has been within the fifty states, where by the end of the Cold War the Department of Defense directly controlled over ten million hectares of land (close to twenty-five million acres) and leased another 80,000 hectares (more than 197,000 acres) from other federal agencies. The Department of Energy, which is responsible for the production of nuclear weapons, also controlled around 10,000 hectares (approximately 24,710 acres). Many of the same problems that existed at bases overseas affected the environment within the United States. In 1993 the Department of Defense listed more than 19,000 contaminated sites in over 1,700 active military facilities across the country. Although portions of these vast holdings have suffered extensive and lasting damage, some areas have been spared the environmental disruption that accompanies commercial development, their presumable fate had they not been controlled by the military.[24]

Much of the environmental destruction wrought by the military during the twentieth century has been connected to the use of oil-fueled weapons systems and transport vehicles, including naval vessels and aircraft. Although the navies of the United States and the great European powers played a relatively minor role in World War I, oil and the internal combustion engine heralded a revolution in mobility on land, sea, and in the air. Oil was also used in the manufacture of munitions and synthetic rubber.[25]

Although the development of nuclear weapons and ballistic missiles and, more recently, the integration of computer technology into weapons systems have fundamentally altered the nature of warfare, oil remains vital to U.S. military power. Each new generation of weapons has also required more oil than its predecessor. For example, according to one estimate, a U.S. armored division in World War II consumed about 60,000 gallons of gasoline daily.[26] In contrast, a 2001 Defense Science Board study estimates that an armored division requires ten times that amount of fuel per day.[27]

During the Cold War the need to ensure sufficient supplies of oil for U.S. forces in Europe led to the creation of the Central European Pipeline System (CEPS), one of the most complex and extensive pipeline systems in the world. Running from the Atlantic and Mediterranean coasts of France to Heidelberg and Amsterdam, the CEPS included over fifty depots, nearly thirty pumping stations, and close to 6,000 kilometers of buried pipeline.[28] In one case, jet fuel leaked from underground pipelines at the Rhein-Main Air Base in Germany and contaminated the underground aquifer supplying water to the city of Frankfurt.[29] There were also shorter pipelines in other European countries. For example, in the 1950s the United States built a seven-hundred-kilometer-long petroleum pipeline connecting its naval base in Rota, Spain, with airfields at Moron, Torrejon, and Zaragoza because, as one analyst noted, "a wing of B-47s (medium-range bombers) consumes in an afternoon more fuel than the entire Spanish railroad tanker fleet can transport in a month."[30]

TABLE 1.2 ENERGY CONSUMPTION OF SELECTED U.S. MILITARY EQUIPMENT

Equipment	Operating Distance or Time	Fuel Consumption (in liters)
M-1 Abrams Tank, average use	1 kilometer	47
F-15 jet, at peak thrust	1 minute	908
M-1 Abrams Tank, peak rate	1 hour	1,113
F-4 Phantom fighter/bomber	1 hour	6,359
Battleship	1 hour	10,810
B-52 bomber	1 hour	13,671
Non-nuclear aircraft carrier	1 hour	21,300
Carrier battle group	1 day	1,589,700
Armored division (348 tanks)	1 day	2,271,000

Source: Michael Renner, "Assessing the Military's War on the Environment," in *State of the World*, (New York: W. W. Norton, 1991), 137. Courtesy Worldwatch Institute, www.worldwatch.org.

Although the U.S. military shrank somewhat in size following the end of the Cold War, it still maintained more than 150,000 ground vehicles, 22,000 aircraft, and hundreds of oceangoing vessels.[31] This fleet guaranteed that conventional military operations would be an oil-intensive activity (see table 1.2). Indeed, maintaining access to oil has become a top priority for the U.S. military. It is currently the single largest consumer of petroleum products in the world and a major contributor to global carbon emissions. Even excluding the additional fuel demands of the Gulf War (1991) and the Serbia-Kosovo bombing campaigns (1999), the U.S. military's annual energy consumption during the 1990s was greater than the total commercial energy consumption of nearly two-thirds of the world's countries.[32] For fiscal year 2003 the Defense Energy Support Center—the agency that manages the military's oil supply—made arrangements to purchase around 184 million barrels of petroleum products. Jet fuels made up 73.5 percent of the total, followed by ground fuels (18.6 percent) and marine fuels (7.93 percent).[33]

According to one estimate, the U.S. military at the end of the Cold War accounted directly for around 2 to 3 percent of total U.S. energy demand and 3 to 4 percent of oil demand, including almost 27 percent of the nation's total consumption of jet fuel. These figures do not include the energy used in manufacturing weapons. Moreover, the military's share of total U.S. use of nonfuel minerals ranged from 5 to 15 percent, and mining operations to extract these minerals often caused significant environmental damage.[34]

In addition to the direct impact of military activities on the environment, numerous indirect effects included new economic and settlement patterns, higher energy consumption, and a greater reliance on private transit. During World War II, U.S. defense spending determined the location of much of the nation's industrial activity and thus had a major influence on its economic geography and population patterns. The U.S. military budget fell between 1945 and 1950 but rose sharply during the Korean War, reaching 12.7 percent of the Gross National Product (GNP) in 1954 and averaging around 5 percent of the GNP from 1950 through 1990. As a result, highly specialized enclaves devoted to military production arose in California, Colorado, the Pacific Northwest, New England, and the South, a development that hastened the decline of the old industrial heartland stretching from New York to Michigan. These economic and demographic shifts left a significant human footprint in terms of roads, buildings, and water use in the South and West.[35]

National-security arguments provided political support for the U.S. Interstate Highway System, first conceived in 1944 and approved by Congress in 1956. Between 1956 and 1970 the federal government spent around $70 billion on the 68,000-kilometer (42,000-mile) National System of Interstate and Defense Highways linking the country's major population centers. During the same period the federal government spent $795,000 on rail transit.[36] The patterns of social and economic organization fostered by the U.S. military have resulted in the subsidization of private automobile use, the consequent deterioration of public transportation, and continuing population shifts to the suburbs as well as to the South and West, changes that have stimulated ever-higher levels of energy use. These have contributed greatly to carbon dioxide buildup, the proliferation of ambient lead (before lead was eliminated from gasoline in the United States in the 1980s), and other atmospheric pollution problems deriving from vehicle use.

Easily the most durable environmental effects of the American military will be those associated with the atomic weapons program: one of the radionuclides generated, plutonium-239, has a half-life of up to 480,000 years. In the United States, as elsewhere, the military built nuclear weapons to address immediate threats and gave scant consideration to the long-term consequences. The American program began in 1942 and by 1945 had bombs ready for use, two of which were dropped on Japan, bringing World War II to a close. Then, in the context of Cold War anxieties, the Americans expanded their nuclear arsenal and tested over a thousand weapons before agreeing to a moratorium on testing in 1993.

The tests, which involved deliberate releases of radiation into the environment, contaminated almost every corner of the country and the globe (although it is impossible to separate the effects of American atmospheric testing from those of the USSR, Britain, and France, which together tested roughly as many weapons as did the Americans). Most testing took place in sparsely populated desert regions (often on Shoshone Indian lands in Nevada), although the United States also tested hydrogen bombs on two Pacific atolls, leaving them devastated by blasts and dangerously contaminated for the foreseeable future. Testing killed a few thousand sheep in Nevada and probably killed a few thousand people indirectly, through cancers resulting from radiation exposure. But the epidemiology of cancer and its relation to radiation remains controversial, and little reliable data exists.

The longer-term effects are linked to the problem of nuclear waste. The atomic weapons program generated radioactive waste on thousands of sites across the country, most notably the bomb factories of the Hanford Engineering Works (in Washington State) and the Rocky Flats Arsenal (in Colorado). In the heat of Cold War tensions, the issue of waste management was left to future generations to resolve, so that containment of radioactive materials, of which there are millions of tons, was at first (until the 1970s) rather haphazard. Groundwater contamination is among the most immediate risks, especially around Hanford. Since the 1970s, and especially since the 1990s, the Department of Defense has undertaken cleanup efforts that ultimately will cost far more than was spent to build the weapons in question. But much of the mess can never be cleaned up: the United States, like other nuclear powers, blithely assumed a radioactive waste management obligation that will last tens of thousands, perhaps hundreds of thousands of years, a period of time far longer than states, civilizations, and even (in the case of plutonium-239) anatomically modern human beings have existed. Future historians will face a daunting task when they seek to explain to generations yet unborn the mentality that led authorities in the United States, and elsewhere, to create this burden.[37]

Combat and the Environment since 1941

In comparison to the environmental effects of preparation for war, those of combat itself appear modest and fleeting. Since 1941, the U.S. military has fought a half-dozen genuine wars and involved itself in several additional smaller operations. We will briefly consider three of these conflicts: World War II, the Vietnam War, and the 1991 Gulf War.

In World War II the main theaters of operations for U.S. forces were North Africa, southern and western Europe, and the Pacific. The ground war in North Africa and Europe produced initially devastating environmental consequences, especially in cities, but that damage was repaired by patient labor within ten to twenty years. Tank warfare in the North African desert, which broke the sand crusts, led to more intense sandstorms for many decades.[38] Some of the Pacific islands that saw bitter fighting—Saipan, Iwo Jima, and Okinawa, for example—were almost entirely shorn of vegetation. But once combat ceased, patterns of ecological succession took hold, approximately as they do after a natural fire. On many atolls, U.S. forces built airstrip pavements out of coral, which destroyed local reefs at least temporarily. In some respects World War II combat brought a respite from ordinary peacetime environmental pressures. For example, the dangers of submarine warfare obliged fishing fleets to sit out the war in port, giving North Atlantic fisheries four years in which to flourish. (Oil spills from sinking tankers and depth charges posed a far smaller risk for fish than did fishermen.)

In Vietnam, where the Americans took over a colonial war from the French in the 1950s and committed forces heavily after 1964, the major environmental effects came from aerial attacks. American bombers put about twenty million craters in Vietnam (many of which now serve as fishponds), and the United States used chemical defoliants and herbicides on a large scale (a practice pioneered by the British in the Malaya insurgency of the 1950s) in attempts to remove the forest cover so vital to guerilla operations. This reduced the forest area of Vietnam by about 23 percent.[39] The defoliants also caused severe health problems for thousands of Vietnamese and for many American soldiers as well.[40]

In the case of the Gulf War, by far the most dramatic environmental damage was perpetrated in Kuwait by Iraqi forces, who ignited some seven hundred oil wells, causing fires that blackened the skies for months and lowered surface temperatures by about ten degrees centigrade. Intentional spills generated temporary oil rivers and lakes in Kuwait and contaminated 40 percent of the country's water supply, polluting it so thoroughly that it remains unusable. The spills also damaged the shorelines of the Persian Gulf for hundreds of kilometers, although with few apparent long-term effects—which in any case are now hard to distinguish from the harm caused by the oil spills that routinely occur in the Gulf and by earlier spills associated with the Iran–Iraq War of the 1980s.[41] The Iraqis also laid more than 1.5 million landmines in Kuwait. One way or another, almost a third of the surface area of the country was affected. The damage to desert and marine life was substantial. Fortunately Kuwait is a rich country and can afford to pay for environmental remediation—although there is none to be had at any price for the contaminated groundwater and some of the oil-soaked soils.[42]

For their part, the Americans sank numerous Iraqi oil tankers and bombed one oil terminal, adding to the marine spills. They also used armor-piercing shells of depleted uranium, which, upon hitting their targets (or anything else), released uranium oxide into the air. The health effects of this type of ordnance are much disputed, but it may be responsible for birth defects and pediatric cancers.[43] Iraq is not in a position to pay for much environmental remediation and has in any case other priorities—including, since 2003, another war that will surely leave its own environmental legacy.

Conclusion

Before the 1890s the environmental implications of the U.S. military hardly reached beyond the nation's borders, and its chief mission was to support settlement with all the ecological changes that process implied. But from the 1890s, the United States sought a larger role in world politics, and the ecological effects of its military establishment became increasingly global. That process began with the acquisition of overseas bases after 1898 and culminated in the post-1941 sprawling archipelago of bases and installations around the world.

The direct environmental impacts were mainly those associated with military bases at home and abroad, with their infrastructure, with their chemical and nuclear wastes, and with the disruptions caused by training and maneuvers—all of which was subject to minimal regulation, especially overseas. The military's fuel use alone was staggering; an F-16 fighter jet burned as much fuel in an hour as the average American motorist used in two years. Impacts even extended into space, because the American military was among the earliest and most prolific producers of "space junk," which includes discarded pieces of satellites, rocket boosters, and the like. Combat, too, had its effects on the environment, particularly in labile ecosystems such as those of deserts or small islands.

Beyond these direct effects, the military contributed to new efficiencies in American manufacturing in the nineteenth century, to new industries (chemical herbicides and pesticides) in the early twentieth century, to new settlement patterns in the United States after 1941, and to a way of life that involved unprecedented levels of energy use. The United States was of course not alone in recasting ecology at home and abroad while preparing for and fighting wars. It was unique only in the extent to which it did so, an extent that corresponded to the global reach of American military power over the past sixty years. The ambition to dominate world politics seemed to require that American society, not merely the U.S. military, seek to dominate nature as well.

Notes

1. For orientation within the meager literature on environmental dimensions of war, a good place to start is Richard P. Tucker and Edmund Russell, eds., *Natural Enemy, Natural Ally: Toward an Environmental History of War* (Corvallis: Oregon State University Press, 2004).

2. Murray Feshbach and Alfred Friendly Jr., *Ecocide in the USSR: Health and Nature under Siege* (New York: Basic Books, 1992); Bellona Foundation, www.bellona.no, a Norwegian-based Web site devoted largely to Russian and Soviet misadventures.

3. Andrew Isenberg, *The Destruction of the Bison* (New York: Cambridge University Press, 2000).

4. Alfred W. Crosby, *Ecological Imperialism: The Biological Expansion of Europe, 900–1900* (New York: Cambridge University Press, 1986).

5. Todd Shallat, *Structures in the Stream: Water, Science, and the Rise of the U.S. Army Corps of Engineers* (Austin: University of Texas Press, 1994), 202.

6. David Hounshell, *From the American System to Mass Production, 1800–1932: The Development of Manufacturing Technology in the United States* (Baltimore: Johns Hopkins University Press, 1984).

7. We thank Peter Shulman of Case Western Reserve University for directing us to this issue and these data, which come from John G. Clark, *Energy and the Federal Government: Fossil Fuel Policies, 1900–1946* (Urbana: University of Illinois Press, 1987), 4, 8–9; and *An-*

nual *Reports of the Navy Department for the Year 1900. Report of the Secretary of the Navy: Miscellaneous Reports* (Washington, D.C.: Government Printing Office, 1900), 285; *Annual Reports of the Navy Department for the Fiscal Year 1919* (Washington, DC: Government Printing Office, 1920), 104.

8. For information on the evolution of the U.S. base system, see C. T. Sandars, *America's Overseas Garrisons: The Leasehold Empire* (Oxford: Oxford University Press, 2000), 27–31.

9. See Richard P. Tucker, *Insatiable Appetite: The United States and the Ecological Degradation of the Tropical World* (Berkeley: University of California Press, 2000).

10. Quoted in Walter LaFeber, *The American Age: U.S. Foreign Policy at Home and Abroad, 1750 to the Present* (New York: W. W. Norton, 1994), 293.

11. Edmund Russell, *War and Nature: Fighting Humans and Insects with Chemicals from World War I to Silent Spring* (New York: Cambridge University Press, 2001).

12. See Jeffrey Legro, *Cooperation under Fire: Anglo-German Restraint during World War II* (Ithaca, N.Y.: Cornell University Press, 1995).

13. See David E. Omissi, *Air Power and Colonial Control: The Royal Air Force, 1919–1939* (Manchester, U.K.: Manchester University Press, 1990).

14. Ibid.

15. John Opie, *Nature's Nation: An Environmental History of the United States* (Ft. Worth, Harcourt Brace, 1998).

16. Melvyn P. Leffler, *A Preponderance of Power: National Security, the Truman Administration, and the Cold War* (Stanford, Cal.: Stanford University Press, 1992).

17. In addition to Sandars, *America's Overseas Garrisons,* see Joseph Gerson and Bruce Bichard, eds., *The Sun Never Sets: Confronting the Network of U.S. Military Bases* (Boston: Beacon Press, 1991); Chalmers Johnson, *The Sorrows of Empire: Militarism, Secrecy, and the End of the Republic* (New York: Metropolitan Books, 2004).

18. U.S. General Accounting Office, *Hazardous Waste: Management Problems Continue at Overseas Military Bases,* Report to the Chairman, Subcommittee on Environment, Energy and Natural Resources, Committee on Government Operations, House of Representatives, GAO/NSIAD-91-231 (Washington, D.C.: Government Printing Office, 1991), 8 (hereafter GAO, *Hazardous Wastes*); John M. Broder, "U.S. Military Leaves Toxic Trail Overseas," *Los Angeles Times,* June 18, 1990.

19. GAO, *Hazardous Wastes,* 33. The unclassified version of the study released by the GAO deleted the names of the bases and the countries where they were located.

20. Michael Renner, "Assessing the Military's War on the Environment," in *State of the World, 1991,* ed. Lester R. Brown et al. (New York: W. W. Norton, 1991), 133, 135; Stephen Dycus, *National Defense and the Environment* (Hanover, N.H.: University Press of New England, 1996), 54, 80.

21. Adm. Eugene Carroll, "U.S. Military Bases and the Environment: A Time for Responsibility," address to the First International Conference on U.S. Military Toxins and Bases Clean-up, Manila, November 23–26, 1997, http://www.yonip.com/main/articles/responsibility.html, accessed October 11, 2007. See also Richard A. Wegman and Harold G. Bailey Jr., "The Challenge of Cleaning Up Military Waste When U.S. Bases Are Closed," *Ecology Law Quarterly* 21 (1994): 937–38; Benjamin Pimental, "Deadly Legacy, Dangerous Ground: Leftover Bombs, Chemicals Wreak Havoc at Former U.S. Bases in Philippines," *San Francisco Chronicle,* July 5, 2001.

22. John Lindsay-Poland, *Emperors in the Jungle: The Hidden History of the U.S. in Panama* (Durham, N.C.: Duke University Press, 2003), 51–55, 61–73.

23. Ibid., 61, 139.

24. Renner, "Assessing the Military's War on the Environment," 134; Dycus, *National Defense and the Environment*, 5, 80; Seth Shulman, *The Threat at Home: Confronting the Toxic Legacy of the U.S. Military* (Boston: Beacon Press, 1992).

25. David S. Painter, "Oil," in *Encyclopedia of American Foreign Policy: Studies of the Principal Movements and Ideas*, 2nd ed., ed. Alexander DeConde, Richard Dean Burns, and Fredrik Logevall, 3 vols. (New York: Charles Scribners' Sons, 2002), 3:1–20.

26. Robert Goralski and Russell W. Freeburg, *Oil and War: How the Deadly Struggle for Fuel in WW II Meant Victory or Defeat* (New York: William Morrow, 1987), 167.

27. U.S. Department of Defense, Office of the Under Secretary of Defense for Acquisition, Technology, and Logistics, *More Capable Warfighting through Reduced Fuel Burden: Report of the Defense Science Board Task Force on Improving Fuel Efficiency of Weapons Platforms*, http://www.acq.osd.mil/dsb/reports/fuel.pdf (2001), p. 13, accessed October 6, 2007.

28. Simon Duke, *United States Military Forces and Installations in Europe* (Oxford: Oxford University Press, 1989), 352–56.

29. Shulman, *Threat at Home*, 110.

30. Quoted in Sandars, *America's Overseas Garrisons*, 151.

31. U.S. Department of Defense, *More Capable Warfighting through Reduced Fuel Burden . . .* , p. 3, accessed October 11, 2007.

32. Vaclav Smil, *Energy at the Crossroads: Global Perspectives and Uncertainties* (Cambridge, Mass.: MIT Press, 2003), 81.

33. U.S. Department of Defense, Defense Energy Support Center, *Energy Support for Global Missions: Fact Book FY 2003*, 56.

34. Renner, "Assessing the Military's War on the Environment," 137–40. The U.S. military also used large amounts of ozone-depleting substances such as halons and chlorofluorocarbons.

35. Ann Markusen, Peter Hall, Scott Campbell, and Sabina Deitrick, *The Rise of the Gunbelt: The Military Revamping of Industrial America* (New York: Oxford University Press, 1991).

36. Steven A. Schneider, *The Oil Price Revolution* (Baltimore: Johns Hopkins University Press, 1983), 60.

37. Arjun Makhijani, Stephen I. Schwartz, and William J. Weida, "Nuclear Waste Management and Environmental Remediation," in *Atomic Audit: The Costs and Consequences of U.S. Nuclear Weapons since 1940*, ed. Stephen I. Schwartz (Washington, D.C.: Brookings Institution Press, 1998), 353–93; J. R. McNeill, *Something New under the Sun: An Environmental History of the Twentieth-Century World* (New York: W. W. Norton, 2000), 342–43; Claus Bernes, *Will Time Heal Every Wound? The Environmental Legacy of Human Activities* (Stockholm: Swedish Environmental Protection Agency, 2001), 147; Len Ackland, *Making a Real Killing: Rocky Flats and the Nuclear West* (Albuquerque: University of New Mexico Press, 2002); U.S. Department of Defense, "Departmental Issues Annual Environmental Cleanup Report for Fiscal 2001," press release, April 11, 2002.

38. M. S. El-Shobosky and Y. G. Al-Saedi, "The Impact of the Gulf War on the Arabian Environment," *Atmospheric Environment* 27A (1993): 95–108.

39. Rodolphe de Koninck, *Deforestation in Viet Nam* (Ottawa: International Development Research Centre, 1999), 12.

40. Arthur P. Westing, *Warfare in a Fragile World: Military Impact on the Human Environment* (London: Taylor and Francis, 1980); Arthur P. Westing, *Environmental Hazards of War* (Newbury Park, CA: Sage Publishers, 1990).

41. Bertrand Charrier, "Human and Ecological Consequences of the Gulf War's Environmental Damages in Kuwait," www.gci.ch/GreenCrossPrograms/legacy/UNCCKUWAIT.html (Geneva: UN Compensation Commission, 2000).

42. *EcoCompass*, "The Environmental Impacts of War," www.islandpress.org/eco-compass/war/war.html (2002); Samira A. S. Omar, E. Briskey, R. Misak, and A. A. S. O. Asem, "The Gulf War Impact on Terrestrial Environment of Kuwait: An Overview," paper presented to the First International Conference on Addressing Environmental Consequences of War, organized by the Environmental Law Institute and held at the Smithsonian Institution, Washington, DC, in 1998.

43. Robert Fisk, cited in *EcoCompass*, "The Environmental Impacts of War."

CHAPTER 2

WOOD FOR WAR

The Legacy of Human Conflict on the Forests of the Philippines, 1565–1946

Greg Bankoff

In this age of plastic, concrete, and steel, it is all too easy to forget how ubiquitous a material wood was in the past. Trees were used to produce shelter, transportation, furniture, utensils, paper and ink, medicines, heat, and even clothing. The tools of agriculture—the plow and dibbling stick—were primarily wooden, as were most weapons of war: the shafts of arrows and spears, the hilts of swords, the palisades of forts, and the hulls of canoes and warships. In tropical landmasses the use of wood was even more commonplace, as the sheer extent of the forests as well as the variety, size, and shape of its trees precluded the use of alternatives except in the case of ostentatious display or absolute necessity. All this wood initially came from the forest, not from plantations; so the history of the forest is largely commensurate with the history of the societies that lived in and about it.[1]

The formation of complex societies always exacts a "price" from the forest and is usually associated with the development of "core" regions and the emergence of state systems.[2] Because formation of a core region in the Philippines occurred after 1565, later than in any other major region of Southeast Asia, the parallel development of the state and the deforestation of the environment are more readily traceable and more extensively documented than elsewhere. On the one hand, this is a narrative of the construction of urban and municipal centers as sites of administration or evangelization and, on the other, of the development of an early agricultural market and the introduction of new crops mainly from the Americas. The maintenance of this state so far from Europe also necessitated its defense against enemies both from without and within, and the construction of ships and forts to protect it inevitably led to exploitation of the islands' timber resources.

However, simply equating state formation and development with deforestation is too crude an analysis. Although, ultimately, the increase of one was mainly at the expense of the other, the specific woods that were required determined which species of trees were disproportionately felled. While the tropical forests of Southeast Asia are noted for their particularly rich flora and fauna, not all wood serves human purposes equally well.[3] Besides, the demand for timber has become far less discerning, especially during the twentieth century. Prior to that time, particular woods were selected for specific purposes, the others often being viewed as worthless. Many tropical hardwoods, such as Southeast Asian teak (*Tectona grandis*), proved ideal for shipbuilding, being both extraordinarily durable and naturally resistant to shipworm.[4] So sought-after did this wood become that the Dutch differentiated only teak or *jati* from other tropical hardwoods on Java, the rest of which were dismissed as worthless "junglewood."[5]

While entire areas of forest were gradually cleared for agricultural and settlement purposes, certain woods were actively sought out for their distinctive properties. This was the case with the timber best suited to military purposes, which was selected for its strength, density, and elasticity. While such selection contributed to the gradual deforestation of an area over time, it also caused a much more drastic decline in certain species and even in particular age groups within species. As mature specimens became exhausted in an area, trees were eagerly sought elsewhere, with the range limited only by ease of access and means of transport, largely fluvial in the preindustrial age. Some species were therefore felled at a much faster rate than the forest in general. This loss of genetic material is known as "erosion" and may be defined as a permanent reduction in the richness or evenness of common localized forms (known as alleles) or the loss of a combination of alleles over time within a defined area. Genetic erosion is detrimental not only to the short-term viability of individuals and populations but also to the evolutionary potential of species.[6] Too few individuals of a species threaten its long-term genetic integrity, increasing the likelihood of chance extinctions and aggravating the effects of inbreeding depression or the size and vitality of future generations.[7]

In the Philippines the demand since the late sixteenth century for certain types of timber preferred for the construction of buildings, fortifications, and ships has severely reduced the supply of specific valuable hardwoods, leaving existing stands isolated and making them more subject to the effects of genetic erosion.[8] While it is unclear whether any species has become extinct as yet, some trees have become exceedingly rare and older, larger breeding specimens almost nonexistent.

This chapter traces the outline of human agency and its effects on the forests of the Philippines in the context of the demands of warfare, examining three distinct military enterprises at different periods: shipbuilding as a response to the threat of Dutch and Moro raiders during the seventeenth and eighteenth centuries, construction of forts in an attempt to subjugate the Cordilleras region of Luzon during the nineteenth century, and the agricultural autarky imposed by Japanese occupiers between 1942 and 1945. While warfare was a significant factor in deforesting the Philippines, its effects were not uniform, leading to the gradual genetic erosion of certain species even before the onset of widespread commercial logging in the twentieth century.

Wood for War

At the time of European contact in 1521, most of the archipelago was covered in forest, though the extent to which timber was a factor in determining Spanish interest is difficult to gauge.[9] Certainly early visitors were struck by its abundance and variety. Although the colonial discourse remains predominantly one of profusion and plenitude throughout the seventeenth and eighteenth centuries, complaints about the scarcity of timber appear as early as 1680.[10] It is impossible, however, to form a reliable estimate of forest cover across the archipelago until the latter half of the nineteenth century, when the deforestation of parts of Luzon was already considerable. One of the chief architects of tropical forestry in the Philippines, Sebastián Vidal y Soler, records how the destruction wrought by "axe and fire has no moment of repose" and concludes that the "myth" of an archipelago covered in an inexhaustible mantle of tropical forest "could not be further from the truth."[11] A concentric pattern of exploitative logging with Manila at its center had already been firmly established and was set to escalate during the ensuing American colonial period.

As might be expected, the statistical ratios show an inverse relationship between forest cover and population growth (table 2.1).[12] According to these calculations, at least 30 percent of the archipelago's woodlands had already disappeared prior to the era of commercial logging, and the scale of this destruction was instrumental in the creation of a forestry department, the Inspección General de Montes, in 1863. The subsequent U.S. administration soon became aware of the state of many of the islands' forests. George Ahern, appointed first chief of the Bureau of Forestry in 1900, noted how "much timber has been cut," but primary forest was still estimated to cover more than ten million hectares.[13] The period from the turn of the century to World War II witnessed an enormous expansion of commercial logging, but it was concentrated in certain provinces as the timber frontier radiated farther out from Manila, affecting especially southern Luzon, Negros, and Mindanao. Moreover, valuable hardwoods were felled in greater quantities than other species—trees selectively cut from which to carve statues, make furniture, construct buildings, and, of course, wage war.

Warfare was an important and highly developed aspect of daily life in the pre-Hispanic Philippines.[14] Weaponry was mainly for hand-to-hand combat and consisted of daggers, knives, swords, spears, javelins, shields, and body armor. Archery was practiced but not widespread, and firearms were restricted to the Muslim south. Most of these weapons were made, at least in part, of wood: the hilts of bladed weapons, the shafts of fighting spears, the fire-hardened heads of bamboo javelins and arrows, the breastplates made of

TABLE 2.1 FOREST COVER AND POPULATION, 1565–1948[15]

Year	Forest cover (in hectares)	Percentage of total area covered	Population
1565	27,500,000[a]	92	800,000[b]
1875	19,405,915[c]	65	6,173,632[d]
1903	20,740,720[e]	69	7,635,426[f]
1918	18,819,281[f]	63	10,314,310[f]
1932	16,950,873[f]	57	13,636,350[f]
1948	17,495,192[g]	58	19,144,000[h]

Sources: (a) *The State of the Philippine Environment* (Manila: Ibon Foundation, 2000), 2; (b) Laura Junker, *Raiding, Trading, and Feasting: The Political Economy of Philippine Chiefdoms* (Honolulu: University of Hawai'i Press, 1999), 62. A higher figure of 1,250,000 people is given in Onofre Corpuz, *The Roots of the Filipino Nation*, 2 vols. (Quezon City: Aklahi Foundation, 1989), 1:29; (c) Sebastián Vidal y Soler, *Memoria de la Colección de Productos Forestales Presentada por la Inspección General de Montes de Filipinas en la Exposición Universal de Filadelfia* (Manila, 1875), 40; (d) José Montero y Vidal, *El Archipiélago Filipino y las Islas Marianas, Carolinas y Palaos: Su Historia, Geografía y Estadística* (Madrid, 1886), 156–60, data for 1877; (e) *Census of the Philippine Islands, 1903*, 4 vols. (Washington, D.C.: GPO, 1905), 1:77; (f) *Annual Report of the Director of Forestry of the Philippine Islands, 1902–1936* (Manila, 1933), 676, 732–37; (g) Florencio Tamesis, "Philippine Forests and Forestry," *Unasylva* 2, no. 6 (1948): 316–25, esp. 316; (h) Jan Lahmeyer, *Population Statistics: The Philippines, Historical Demographical Data of the Whole Country*, http://www.populstat.info/Asia/philippc.htm.

hardwood, and the shields constructed from fibrous, corky woods.[16] As inhabitants of an archipelago, Filipinos spent much of their time on water, and many of their military engagements took place at sea; Philippine technology in the construction of warships was therefore highly developed. *Caracoas* were sleek, double-prowed vessels of low freeboard and shallow draft that mounted a square sail. They had double outriggers on which up to four banks of paddlers could provide speed under battle conditions and a raised platform amidships for warriors. They were light, flexible, extremely maneuverable, and perfectly suited to the maritime conditions in which they operated. And as Europeans learned to their discomfort, they were fighting machines par excellence, able to reach speeds of between twelve and fifteen knots, in contrast to the five or six knots made by a galleon.[17] "Their ships sail like birds," commented Francisco Combés in 1667, "while ours are like lead in comparison."[18] The prevalence of these crafts and thus the amount of wood felled for their construction is impossible to ascertain, although the Maguindanaos were estimated to have mustered a war fleet of a hundred such vessels in 1602, and Rajah Bongsu of Jolo apparently set out with as many as two thousand fighting men in 1627.[19] These and similar reports indicate that this type of craft was both common and numerous. Despite the fact that warfare was endemic in precontact societies, however, its environmental effect was likely to have been limited and localized given the weaponry available and the relatively low population density.[20]

Shipbuilding in the Seventeenth and Eighteenth Centuries

All this changed in the aftermath of the 1565 Spanish colonization and the transference to the Philippines of both European rivalries and religious antagonisms. These new conditions took the form of defending the colony against Dutch fleets and Moro (Muslim) raiders. On the one hand, the Spanish presence dragged the archipelago into a wider arena of conflict: the Eighty Years' War between Spain and the Netherlands that was finally settled only in 1648 by the Treaty of Westphalia, which formally recognized the seven Dutch provinces as independent. On the other hand, it bestowed a new ferocity on the slave-raiding expeditions out of Jolo and Maguindanao. While slave raiding certainly predates 1565, it intensified as a result of the religious tensions and settlement policies introduced by the Spanish colonizers. Dutch activities in Philippine waters were organized by the Dutch East India Company, or Vereenigde Oostindische Compagnie (VOC), from their bases in the Moluccas, Java, Formosa (Taiwan), and Japan. The VOC's aims were largely commercial: to hinder trade in the north by blockading Manila and intercepting the bullion-loaded Acapulco galleons, and, in the south, to thwart Spanish attempts at establishing a permanent presence in the Spice Islands. During the first quarter of the seventeenth century, the VOC dispatched no fewer than sixteen fleets, fought four major naval battles (1601, 1616, 1617, and 1626), and maintained from forty to fifty armed vessels in Philippine waters during any one year.[21] War with the Muslims was aggravated by Spanish attempts to subjugate these polities in the south and by a colonial policy that concentrated people and wealth in coastal cities. Hostilities took the form mainly of Moro raiding and Spanish retaliatory expeditions, which subsided only after the death of Sultan Kudarat of Magindanao in 1671.[22] The scale of these campaigns was considerable, and the losses incurred caused an endless drain on both labor and resources. Nor did the Spaniards always have the best of it; there were an estimated ten thousand Christians held captive in 1621, and the colonial authorities were forced to abandon their forts in the south in 1662.[23] For most of the eighteenth century, the Philippines were effectively divided between the Muslim sultanates of Magindanao and Sulu in the south and a Hispanicized Christian realm in the rest of the archipelago.[24]

The Spanish ships needed to defend their new possessions in the East; they were therefore quick to appreciate the Philippines' potential to supply that need, as their early accounts of the islands show.[25] In a letter to Philip II, Juan Maldonado estimated that there was enough timber in the archipelago to build three to four galleons each year, while Francisco Leandro de Viana reported that there was enough for "at least ten."[26] Not only was there an "abundance of wood," but Filipinos also proved to be "very skilful in making ships" and could do so "quickly and cheaply."[27] And build ships is what the Spanish did. The first shipyards were established at Cavite and Oton, but vessels were also constructed in Masbate, Marinduque, Camarines, and Albay. By 1616, six out of the seven galleons stationed at Manila had been built in the islands.[28] These were not just smaller crafts and galleys but large ships: the *Santa Rosa*, begun in 1674, was one of the finest ships of its age; while the *San Jose*, launched in 1694, was reputedly the tallest ship afloat anywhere in the world.[29] At the same time, however, there was a continual need to refurbish or replace these vessels due to wartime damage and losses, to the relatively rapid decay of timbers in tropical waters, and to frequent shipwrecks.[30]

The quality of Philippine timber was ideal for ship construction. Molave (*Vitex parviflora*) was the principal wood used for futtock-timbers and stem-crooks; guijo (*Shores guiso*), yacal (*Hopea malibato*), betis (*Ganua manticola*), dungon (*Heritiera littoralis*), and ipil (*Leucana leococephala*) were preferred for keels and stern-posts; banaba (*Lagerstroema speciosa*) for exterior planking, as it did not rot and was resistant to shipworm; lauan (*Shorea contorta*) and tanguile (*Shorea polysperma*) for the hull planking, as these woods did not chip when struck by cannonballs and reputedly absorbed much of the impact; mangachapuy (*Hopea accuminata*) for masts because of its elasticity; and palo maria (*Calophyllum inophyllum*) for yards and topmasts.[31] Other woods were used for more specialized purposes.[32] If properly seasoned, timbers could withstand the sea and the elements for fifty or sixty years. Ramón Jordana y Morera makes mention of two brigantine schooners, the *Soledad* and the *Feliz Esperanza*, constructed in Pangasinan in 1825–26 and still in active service in 1877.[33] But vessels were often built with such haste that unseasoned wood was used, which meant that "one must tear up the decks every two years and put down new ones."[34] How much timber was consumed in this frenzied construction over the decades is difficult to estimate, but there are indications that it was considerable.[35] Some idea of the scale of the whole enterprise can be gleaned from the labor that was levied to meet these needs. As part of the corvée that indigenous people were forced to render the colonial state, municipalities had to provide people to work in the shipyards or fell timber, the dreaded *corte de madera*.[36] The manpower needed was immense; the masts of one galleon reputedly involved the efforts of six thousand Filipinos for three months simply to transport them.[37] As most of the labor was drawn from the five provinces closest to Manila, most of the timber was cut there, though forests elsewhere in the islands were also affected.[38] In fact, vessels were purposely constructed at sites all around the archipelago to lessen the hardship on the population nearest to the capital and to ensure an adequate pool of labor. So great were the exactions levied on locals that it caused insurrections on more than one occasion; revolts broke out, notably in 1614 and again in 1649, affecting Samar, Leyte, and the Bicol region of southern Luzon.[39]

Forests and, more particularly, certain species of trees in the vicinity of shipyards were felled and their timber used in ship construction as well as for other purposes. By 1618, Alonso Fajardo de Tenza was already lamenting the impossibility of building a fleet to defend the colony, as his predecessor, Juan de Silva, had been able to do, because the latter had "impoverished the wealth ... of the wretched natives to such an extent that many are now in the most dire need."[40] José Rizal's marginalia in a 1609 copy of Antonio Morga's *Sucesos de las Islas Filipinas* (1609) alludes to the apparent disappearance or scarcity of some species of trees, such as betis, because of the excessive demands of shipbuilding.[41] It seems safe to assume that molave and other valuable hardwoods used for similar purposes were also becoming increasingly difficult to find. Already by 1621, Hernando de los Ríos Coronel hints as much in his memorial addressed to the king that talks about "the great difficulties" in locating adequate stands of timber and the need to penetrate "the thicker recesses of the woods."[42] Moreover, much timber was wasted, and the officials in charge of shipyards or organizing the *corte de madera* were notorious for their avarice and malfeasance, which resulted in the felling of many more trees than was strictly necessary, especially those species sought after for their remarkable properties and size.[43] The building of ships continued during the eighteenth century. While the Dutch threat receded, that of the Moro raiders persisted and finally abated only in the mid-nineteenth century in the face of the overwhelming firepower and speed of the steam-driven gunboat.[44]

Fort Construction and the Inland Frontier during the Nineteenth Century

Ships were not the only drain on timber for military purposes. Spanish control over the archipelago was tenuous in many areas and remained so right into the nineteenth century. While it is accurate to talk about the presence of a Christian, Hispanicized society in the lowlands of the northern and central islands, most of Mindanao and even extensive mountainous areas of Luzon lay beyond the effective reach of the colonial administration and were at best occupied only militarily by Spain. The peoples of Mindanao resisted the imposition of Spanish rule and were effectively incorporated into the colonial state only in the early twentieth century during the American period. In the Cordilleras of central Luzon, however, a *presidio* society of forts and military garrisons evolved.[45]

The great chain of mountains known as the Gran Cordillera Central comprises the four contemporary provinces of Apayao-Kalinga, Mountain, Ifugao, and Benguet. The region is home to a number of ethnolinguistic groups that Spaniards indiscriminately labeled *infieles* (pagans).[46] These peoples lived in villages of more-or-less related persons, took heads and slaves when they went to war, and recognized no central authority.[47] Exposed to missionary contact and irregular military expedition since the sixteenth century, they became a serious colonial concern only after the establishment of the Tobacco Monopoly in 1781.⁴⁸ Revenue agents found it impossible to restrict the mountain people's trade in contraband tobacco and responded by establishing a series of military commands to expand their control over the region. Beginning in the 1820s, successive governors-general in Manila tried to impose colonial order and levy tax and labor obligations on its peoples.⁴⁹ The mountains were divided into a number of *comandancias político-militares* and alternately subjected to policies of attraction and repression. The Spanish presence, however, remained primarily a military occupation based on garrison towns and forts, dependent on lowland sources for provisions and soldiers for compliance.⁵⁰

Like the naval craft that defended the colony from the sea, the forts that straddled the highlands were also constructed of timber. Fortifications built by both the indigenous peoples and the Spaniards had long been an aspect of warfare in the archipelago.⁵¹ Often these forts were fairly elaborate affairs, palisades built of "heavy timbers" with parapets protected by moats, earthworks, and outer stockades.⁵² A report to the governor general in 1893 on the timber required to build one such fort, the *Colonia Militar* of Tumauini in Isabela province, gives a fairly detailed account of the amount and type of hardwoods used to construct and maintain such outposts.⁵³ Applying Fred Foxworthy's classifications of their relative strengths, the varieties of timber included: bolongeta (very hard and very heavy), aranga (very hard and heavy), molave (hard and heavy), ipil (hard and heavy), tindalo (hard and heavy), guijo (hard and moderately heavy), and narra (moderately hard and moderately heavy).⁵⁴

Each pueblo or township in Isabela was assigned an amount of timber and a date by which the designated number of beams had to be delivered (figure 2.1).⁵⁵ Initially the governor set this number at 5,658 beams, but he was forced to halve this amount due to the reluctance and/or inability of local authorities to comply with his demands. There were continuous requests from town officials for extensions, including one from the local authorities in Tumauini. There the *gobernadorcillo* (mayor) sought authorization to call out all the town's residents "without distinction or class of *cedula*" to cut wood, as the number of *polistas* available was "insufficient for the required service," but his request was denied.⁵⁶ The twenty-two *polistas* of Tumauini were therefore each obliged to deliver an average of seven to eight beams, three to four *palmas bravas*, twenty-two canes, and two to three bundles of rattan.⁵⁷

Some towns may have had good reasons for failing to meet their quotas. In one case, a flash flood caused by a passing typhoon had washed away much of the timber, which had to be cut anew. Other pueblos simply did not have the *carabaos* (water buffalo) necessary for moving big logs.⁵⁸ In the event, the towns of Isabela delivered 3,139 beams to the colonial military authority at Tumauini, excluding those rejected as substandard by the officer in charge. Given the number of forts situated at strategic locations or important transit points all over the Cordillera, as well as in other parts of the archipelago, the amount of timber required in their construction and maintenance must have been substantial. The wood felled for this purpose was primarily valuable hardwood,

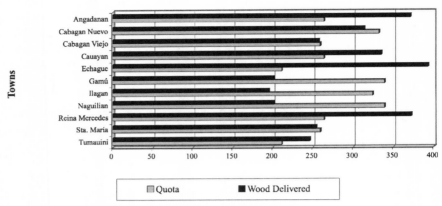

Fig. 2.1 Timber delivered to the Colonia Militar at Tumauini in 1893.

"heavy timbers" such as were used for shipbuilding, further contributing to the reproductive erosion of certain species. In fact, a report summarizing the state of the forest in 1877 blamed the unregulated activities of woodcutters for a scarcity of "valuable trees of large dimensions" and their place being taken by "juvenile ones" which had resulted in the "impoverishment" of the forest.[59] Moreover, there was also a constant need to replace these timbers as a result of the damages caused by *anay* (white ants) that attacked even the hardest woods, especially those portions buried underground.[60] Evidently officials in the Ministerio de Ultramar, the Spanish colonial office in Madrid, were aware of the situation and appreciated that "in Tayabas and other localities of this archipelago woods such as mangachapuy, betis, banaba, and others of the most sought after species ... are already very scarce."[61]

Autarky and the Japanese Occupation, 1941–45

The U.S. occupation of the archipelago in 1899 was also accompanied by warfare that was officially declared over in 1902 but which continued to be waged until 1907 by popular revolutionary leaders who refused to accept the colonial yoke anew. While some woodlands were undoubtedly damaged during the campaigning or ruined as a result of military policies that involved destroying crops and concentrating populations, there appears not to have been any appreciable loss of forest cover according to the rough estimates for the period.[62] The ensuing four decades of American colonial administration witnessed the implementation of many of the reforms and changes in forestry practice that had long been advocated by Spanish forestry experts and led to a steady increase in the amount of timber cut. Mechanized logging practices, the expansion of operations into new areas, and the commercial felling of a wider range of dipterocarps soon transformed the Philippines into the foremost timber exporter in Southeast Asia.[63] Annual lumber production rose from 94,000 to 2,500,000 cubic meters between 1901 and 1940.[64] In particular, the volume of timber cut from lesser-quality dipterocarp hardwoods such as red and white lauan, apitong, and tanguile rose in excess of a thousand percent.[65] Moreover, the demands of a rapidly rising population for new agricultural land also led to substantial forest clearance, especially after the establishment of internal self-rule during the Commonwealth period (1935–41).[66]

The need for both timber and food intensified with the Japanese invasion and the attempt to mobilize the archipelago's forest resources for military purposes. On December 10, 1941, three days after the attack on Pearl Harbor and after a systematic bombing of Philippine defenses, Japanese invasion forces landed simultaneously at several points on Luzon. The Imperial Army rapidly overran the less well equipped and numerically inferior American and Filipino forces, which finally capitulated in May 1942.[67] The military value of timber was recognized by all combatants, but to the Japanese the forests represented an essential component of their overall war strategy.[68] According to Maj. Gen. Yoshihide Hayasi, Director General of the Japanese Military Administration, the occupiers' primary objective was "To develop the [islands'] resources for defense for the purpose of meeting the demands in the Greater East Asia War."[69] The first step toward accomplishing this aim was to impose a policy of autarky—that is, to make the Philippines economically self-sufficient and able to sustain not only native Filipinos but also an occupying force of approximately five hundred thousand Japanese.[70] Filipino labor and the natural resources of the islands—including cotton, sugar, minerals, metals, medicinal substances, and, of course, timber—were all harnessed to the Japanese war effort.

The Japanese were determined to make full use of the archipelago's forest resources in timber and nontimber products. Allied-owned logging and milling plants were expropriated and timber production placed under the control of one large Japanese-run company, the Philippine Lumber Control Union (Philippine Mokuzai Tosei Kumiai), which exercised exclusive operating and buying powers through its subsidiary companies in each province. All operators were obliged to sell their timber to a designated holding company that subsequently supplied the needs of the Imperial Japanese forces. As the demands of the latter were virtually insatiable, very limited amounts of timber were available for nonmilitary purposes.[71] But production remained at a fraction of its prewar levels, despite the strenuous exertions of Japanese officials. Carlos Sulit estimates that monthly production, which in 1941 was just under 184,000 cubic meters, averaged only 8,355 cubic meters between 1942 and 1945; and in the fiscal year immediately following liberation, monthly production jumped to 17,924.[72] The extraordinarily low rate of production during the war years was due to multiple factors: damage to mills from aerial bombing and sabotage, the lack of spare parts to repair them, the progressive shortage of *carabaos* indiscriminately slaughtered to provide food, the activities of guerrilla fighters that confined logging operations to only the most accessible forests, the unrealistically low schedule of timber prices fixed by the Japanese that discouraged production, and the general antipathy of Filipinos for the occupying regime, despite the nominal granting of independence under Jose Laurel on October 14, 1943.[73]

The decimation of the archipelago's forests during the occupation, however, far exceeded the level of destruction reflected in the official production figures or by the unknown amount of timber sold on the black market and cut in areas controlled by guerrilla forces.[74] In the first place, the Japanese invasion precipitated a food shortage. Dependent on imports before the war, the Philippine economy was severely tested by the sudden halt in imports of rice and fish products. Problems were apparent by 1943 and became more acute in each successive year of the occupation. Executive Order No. 40 (May 14, 1942) permitted the cultivation of any "idle" public and private land in an attempt to increase food production and led to an explosion of unregulated slash-and-burn swidden (*kaiñgin*) agriculture and to the clearance of many forest areas.[75] A practice decried by forestry officials since Spanish times, the areas affected by *kaiñgin* expanded rapidly as many people took refuge in mountainous regions. The fate of Makiling National Park in central Luzon is a case in point.[76] Tenants and laborers, whose adjacent fertile rice and sugar fields were commandeered by the Japanese military, cleared approximately two hundred hectares of the park to plant crops. So great was the demand for and so chronic the shortage of timber that trees in national parks and forest reserves were indiscriminately felled.[77] Legally protected species, experimental plantations, special forests, and even fruit trees did not escape the attention of the Japanese; and what was left often fell prey to black marketeers. Moreover, many forest areas were also the scenes of intense fighting, first in Bataan in early 1942 and again in the months following the American landing on October 20, 1944. The sheer scale of the destruction and the random, arbitrary nature of the damage inflicted on the forests separate the Japanese occupation from all previous wartime experience in the Philippines.

Conclusion: War for Wood

The relationship between warfare and environment is as long as the history of complex human societies. The environment in the form of weather, climate, terrain, soil, and vegetation has played an important and sometimes crucial role in combat.[78] But it has also been an unrecognized casualty of war, devastated by increasingly powerful weaponry and now a military target in its own right.[79] Changes in military policy that stress interdiction have escalated environmental destruction, and the deployment of massive firepower against landscapes that cloak the movements of enemies—or against croplands, water supplies, and transportation routes that support their activities—has become standard practice.[80] More important still, the environment is a strategic resource providing the tools and matériel necessary to wage war or sustain populations at war, and so demand for or shortage of various commodities has come to constitute a *casus belli* in itself.[81] According to Thomas Homer-Dixon, as ecological stresses lead to greater social disintegration, which in turn provokes violent conflict, and as that conflict brings about still more environmental degradation, it becomes more and more likely that future wars will be contests for control over scarce natural resources on which communal survival depends.[82]

All these factors are relevant to a greater or lesser extent in the case of the Philippines, but to talk only about the impact of war on the forests does an injustice to the complexity of that relationship.[83] In one sense, certainly, this is a history of how wood for war increasingly became war for wood. It traces how wood was used for weapons in precontact times, for shipbuilding as a defense against the Dutch and Moro threats in the seventeenth and eighteenth centuries, and in fort construction along the Cordillera during the nineteenth century; how then, during the twentieth century, the potential of the archipelago's forests was realized first by the American occupation forces and subsequently, briefly, by the Japanese, transforming wood into a "prize" of war. But the effects of warfare were more selective than such a linear typology might suggest; certain types of valuable hardwoods were required for specific military purposes, and some trees were disproportionately felled in comparison to others. This pattern of overharvesting species for military uses resulted in genetic erosion, a diminishing of particular species' contribution to the general biomass of the archipelago, and the possible degeneration of remaining species through the culling of the largest and most vigorous specimens. The consequences of such activities were larger still, affecting the entire balance of microecosystems and the fauna and flora dependent on them. In fact, the present forests of the Philippines have probably been shaped as much by the selective demands of warfare as they have been by processes of climax formation or natural selection.

Notes

Research presented in this essay was supported in part by a 2003–2004 residential fellowship at the Netherlands Institute for Advanced Study (NIAS) in Wassenaar.

1. Michael Williams, *Deforesting the Earth from Prehistory to Global Crisis* (Chicago: University of Chicago Press, 2003).

2. George Coedès, *The Indianized States of Southeast Asia* (Honolulu: University of Hawai'i Press, 1968); Stanley Tambiah, *World Conqueror and World Renouncer: A Study of Buddhism and Polity in Thailand against a Historical Background* (Cambridge: Cambridge University Press, 1976); and Oliver Wolters, *History, Culture, and Region in Southeast Asian Perspectives* (Singapore: Institute of Southeast Asian Studies, 1982).

3. Chris Barrow, "Environmental Resources," in *South East Asian Development: Geographical Perspectives,* ed. Denis J. Dwyer (Harlow, U.K.: Longman Scientific and Technical, 1990), 93–94.

4. Herbert G. Baker, *Plants and Civilization*, 3rd ed. (Belmont, Cal.: Wadsworth Publishing, 1978), 40.

5. Peter Boomgaard, "Forests and Forestry in Colonial Java, 1677–1942," in *Changing Tropical Forests: Historical Perspectives on Today's Challenges in Asia, Australasia, and Oceania*, ed. John Dargavel, Kay E. Dixon, and Noel Semple (Canberra: Centre for Resource and Environmental Studies, 1988), 63.

6. A. Brown, A. Young, J. Burdon, L. Christides, G. Clarke, D. Coates, and W. Sherwin, *Genetic Indicators for State of the Environment Reporting* (Canberra: Environment Australia, 1997).

7. Timothy C. Whitmore, *An Introduction to Tropical Rain Forests*, 2nd ed. (New York: Oxford University Press, 1998), 232.

8. The term *valuable hardwoods* refers to those hardwood species or groups of hardwood species that are considered to possess special technical properties (strength, natural durability, and good machining properties) and appearance (grain, figure, texture, color, or aesthetic qualities). These valuable hardwoods contrast with lower-quality hardwoods used for more general purposes and to other timbers that serve mainly for fuel or pulp.

9. As a comparison, Peter Boomgaard argues that Dutch acquisition of territory in Java was at least partly motivated by teak stands ("Forests and Forestry," 64–67).

10. See Miguel de Legazpi, "Relation of the Philippine Islands, Cebu, July 7, 1569," in *The Philippine Islands, 1493–1803*, 55 vols., ed. Emma Blair and Alexander Robertson (Cleveland, 1903–1909), (hereafter cited as B&R), 3:54–61, esp. 59; Andrés de Mirandaola, "Letter to Felipe II, January 8, 1574," in B&R, 3:223–29, esp. 225. Pedro Velasco, "Later Augustinian and Dominican Missions, Tondo, April 16, 1760," in B&R, 48:59–136, esp. 91; and Domingo Perez, "Relation of the Zambals, 1680," in B&R, 47:289–332, esp. 292, 293.

11. Sebastián Vidal y Soler, *Memoria sobre el Ramo de Montes en las Islas Filipinas* (Madrid, 1874), 11 and 28.

12. See David Kummer, "Measuring Forest Decline in the Philippines: An Exercise in Historiography," *Forest and Conservation History* 36 (1992): 185–89.

13. George Ahern, *Compilation of Notes on the Most Important Timber Tree Species of the Philippine Islands* (Manila, 1901), 11; and Edwin Schneider, *Commercial Woods of the Philippines: Their Preparation and Uses* (Manila: Bureau of Printing, 1916), 11.

14. William Henry Scott, *Barangay: Sixteenth-Century Philippine Culture and Society* (Quezon City: Ateneo de Manila University Press, 1994), 151.

15. Some of the figures presented in this table are more reliable than others, though all must be regarded as speculative to a certain degree. More reliance can be placed on the 1875, 1918, and 1932 statistics, as these are based on province-by-province assessments. There has been no attempt to reconcile the figures given the contingent nature of their validity; quite clearly, however, it would be absurd to suppose that forest cover increased in 1903 or, again, in 1948. The problems relative to estimating forest cover are discussed at length in Greg Bankoff, "One Island Too Many: Reappraising the Extent of Deforestation in the Philippines Prior to 1946," *Journal of Historical Geography* 33, no. 2 (2007): 314–34. On deforestation since World War II, see David Kummer, *Deforestation in the Postwar Philippines* (Chicago: University of Chicago Press, 1991).

16. Ibid., 147–51.

17. William Henry Scott, *Boat Building and Seamanship in Classic Philippine Society* (Manila: National Museum, 1981), 5–10.

18. Francisco Combés, *Historia de Mindanao y Jolo... Obra Publicada en Madrid en 1667, y que Ahora con la Colaboración del P. Pablo Pastells... Saca Neuvamente á Luz W. E. Retana* (Madrid, 1897), 70–71.

19. Onofre D. Corpuz, *Roots of the Filipino Nation*, 2 vols. (Quezon City: Aklahi Foundation, 1989), 1:141, 143.

20. Laura Lee Junker, *Raiding, Trading, and Feasting: The Political Economy of Philippine Chiefdoms* (Honolulu: University of Hawai'i Press, 1999), 62.

21. Corpuz, *Roots of the Filipino Nation*, 1:123; and Otto van den Muijzenberg, *Four Centuries of Dutch-Philippine Economic Relations, 1600–2000* (Manila: Royal Netherlands Embassy, 2000), 12–15.

22. On Dutch designs on the Philippines, see Ruurdje Laarhoven, *Triumph of Moro Diplomacy: The Maguindanao Sultanate in the 17th Century* (Quezon City: New Day Publishers, 1989).

23. Antonio de Morga, "Sucesos de las Islas Filipinas, Mexico, 1609" (published over two volumes), in B&R, 15:265; and Corpuz, *Roots of the Filipino Nation*, 1:135–58.

24. On Sulu and Maguindanao, see James Francis Warren, *The Sulu Zone, 1768–1898: The Dynamics of External Trade, Slavery, and Ethnicity in the Transformation of a Southeast Asian Maritime State* (Singapore: Singapore University Press, 1981); and James Francis Warren, *Iranun and Balangingi: Globalisation, Maritime Raiding, and the Birth of Ethnicity* (Singapore: Singapore University Press, 2002).

25. Miguel de Legazpi, "Relation of the Philippine Islands, Cebu, July 7, 1569," in B&R, 3:59; Andrés de Mirandaola, "Letter to Felipe II, January 8, 1574," in B&R, 3:225; and Francisco de Sande, "Relation and Description of the Phelipinas [sic] Islands, Manila, June 8, 1577," in B&R, 4:59.

26. Juan Maldonado, "Letter to Felipe II, Manila, 1575," in B&R, 3:303; and Francisco Leandro de Viana, "Memorial of 1765, Manila, February 10, 1765," in B&R, 48:296.

27. Santiago de Vera, "Memorial to the Council by Citizens of the Filipinas Islands, Manila, July 26, 1586," in B&R, 6:206.

28. Sebastián de Pineda, "Philippine Ships and Shipbuilding, Mexico, 1619," in B&R, 18:180.

29. Corpuz, *Roots of the Filipino Nation*, 1:92–93.

30. Pineda, "Philippine Ships and Shipbuilding," 171, 173; Juan de Medina, "History of the Augustinian Order in the Filipinas Islands, 1630," in B&R, 24:85; Sebastian Hurtado de Corcuera, "Letter to Felipe IV, Cavite, July 11, 1636," in B&R, 26:286; and Casimiro Diaz, "The Augustinians in the Philippines, 1641–70, Manila, 1718," in B&R, 37:211.

31. Pineda, "Philippine Ships and Shipbuilding," 169–73; Diaz, "The Augustinians," 251; and Viana, "Memorial," 296.

32. Vidal y Soler, *Memoria*, 144–81.

33. Ramón Jordana y Morera, *Estudio Forestal acerca de la India Inglesa, Java y Flipinas* (Madrid, 1891), 226.

34. Alonso Farjado de Tenza, "Letter to Felipe III, Cavite, August 10, 1618," in B&R, 18:131; and Pineda, "Philippine Ships and Shipbuilding," 173. Unseasoned timbers were also required to facilitate cutting and shaping to the required sizes. On European shipbuilding techniques, see Richard W. Unger, *The Ship in the Medieval Economy, 600–1600* (London: Croom Helm, 1980).

35. Carla Rahn Phillips, *Six Galleons for the King of Spain: Imperial Defense in the Early Seventeenth Century* (Baltimore: Johns Hopkins University Press, 1986), 19–77.

36. Sant Pablo, "Compulsory Service by the Indians, 1620," in B&R, 19:71–76; and Nicholas P. Cushner, *Spain in the Philippines: From Conquest to Revolution* (Quezon City: Ateneo de Manila University, 1971), 117–26.

37. Hernando de los Ríos Coronel, "Memorial y Relacion para su Magestad, Madrid, 1621," in B&R, 19:203.

38. José Vila, "Condition of the Islands Manila, October 7, 1701," in B&R, 44:126.

39. Corpuz, *Roots of the Filipino Nation*, 1:124–28.

40. Tenza, "Letter to Felipe III," 120.

41. B&R, 16:87n71.

42. Ríos Coronel, "Memorial y Relacion," 203.

43. Francisco de Sande, "Relation of the Filipinas Islands, Manila, June 7, 1576," in B&R, 4:115; and José Raón, "Ordinances of Good Governance, 1768," in B&R, 50:246.

44. Warren, *The Sulu Zone*; and Warren, *Iranun and Balangingi*.

45. The Spanish word *presidio* means both a garrison of soldiers and a fortress garrisoned by soldiers. A comparable situation existed on Mexico/New Spain's northern border where a line of garrisoned forts marked the frontier with hostile Apache and Comanche tribes. See Jack Williams, "San Diego Presidio: A Vanished Military Community of Upper California," *Historical Archaeology* 38, no. 3 (2004): 121–34.

46. Six distinct ethnolinguistic groups live in the Gran Cordillera Central region: Isneg (Apayao), Kalinga, Bontoc, Ifugao, Kankanay, and Ibaloy.

47. On local Cordillera societies, see Martin W. Lewis, *Wagering the Land: Ritual, Capital, and Environmental Degradation in the Cordillera of Northern Luzon, 1900–1986* (Berkeley: University of California Press, 1992).

48. On the Spanish colonial government's establishment of a Tobacco Monopoly, see Edilberto C. de Jesus, *The Tobacco Monopoly: Bureaucratic Enterprise and Social Change, 1766–1880* (Quezon City: Ateneo de Manila University Press, 1980).

49. William H. Scott, *The Discovery of the Igorots: Spanish Contacts with the Pagans of Northern Luzon* (Quezon City: New Day Publishers, 1974), 3–4.

50. Ibid., 296.

51. Alonso Beltran, "Expeditions to Borneo, Jolo, and Mindanao, 1576," in B&R, 4:169; and Santiago de Vera, "Letter to Felipe II, Manila, June 26, 1588," in B&R, 7:58.

52. Joseph Fayol, "Affairs in Filipinas, 1644–47, Manila, 1647," in B&R, 35:255; and Scott, *Discovery of the Igorots*, 272–73.

53. Pueblo de Tumauini Copia de los Documentos sobre Corte, Estracción y Entrega de las Maderas Perteneciente a la Colonia Militar No. 70, 1893, Philippine National Archive (hereafter cited as PNA) Corte de Maderas, Bundle 2.

54. Fred Foxworthy, "Philippine Woods," *Philippine Journal of Science* 2, no. 5 (1907): 351–403. These woods were mainly listed by their vernacular names: amaga (bolongeta), cuela (aranga), amugauan (molave), magalayao (tindalo), saray or zilan (guijo).

55. Source: Información sobre las Maderas Entregadas por los Pueblos de Isabela de Luzon a la Colonia Militar de Tumauini, 1893, PNA Corte de Maderas, Bundle 2. The amount of beams in figure 2.1 may also include *palmas bravas* (*Livistona* spp.).

56. A *cedula personal* was an identification paper that denoted the racial classification and tax liability of its bearer; a *polista* was someone engaged in *polo*, or forced labor.

57. Relación Númerico de las Maderas Cortadas por los Vecinos del pueblo de Tumauini y Entregados á la Colonia Militar segun los Recibos Firmados por el Teniente Coronel 1er

Jefe del Regimiento de Linea Magallanes No. 70 que se Expresan á Continuacion, Tumauini, October 22, 1893, PNA Corte de Maderas, Bundle 2.

58. Lopez Nuñez to gobernadorcillo of Tumauini, May 9, 1893, PNA Corte de Maderas, Bundle 2. A great rinderpest epidemic decimated up to 90 percent of *carabao* (water buffalo) herds during the late 1880s. Marshall McLennan, *The Central Luzon Plain: Land and Society on the Inland Frontier* (Quezon City: Alemar-Phoenix Publishing House, 1980), 169; and Reynaldo Ileto, "Hunger in Southern Tagalog, 1897–1898," in *Filipinos and Their Revolution: Event, Discourse, and Historiography*, ed. Reynaldo Ileto (Quezon City: Ateneo de Manila University Press, 1998), 113–15.

59. Memorandum from Ministerio de Ultramar to Governor-general of the Philippine Islands, Madrid, August 7, 1877, PNA Corte de Maderas, Bundle 2.

60. Ahern, *Notes on the Most Important Timber Tree Species*, 91.

61. Memorandum from Ministerio de Ultramar to Governor-general, August 7, 1877.

62. Glenn May, *Battle for Batangas: A Philippine Province at War* (New Haven, Conn.: Yale University Press, 1991), 177, 255–56, 264–65. It was estimated that 70 percent (equivalent to about 20,719,000 hectares) of the archipelago was forested in 1903, while the last Spanish figure put the forest cover at 19,470,600 hectares in 1890 (Census of the Philippine Islands, 1:77).

63. On Spanish policy, see Vidal y Soler, *Memoria* and Greg Bankoff, "Almost an Embarrassment of Riches: Changing Attitudes to the Forests in the Spanish Philippines," in *A History of Natural Resources in Asia: The Wealth of Nature*, ed. Greg Bankoff and Peter Boomgaard (New York: Palgrave Macmillan, 2007), 103–22. On American colonial policy, see Dennis Roth, "Philippine Forests and Forestry: 1565–1920," in *Global Deforestation and the Nineteenth-Century World Economy*, ed. Richard P. Tucker and John Richards (Durham, N.C.: University of North Carolina Press, 1983), 30–49; and Richard P. Tucker, "The Commercial Timber Economy under Two Colonial Regimes in Asia," in *Changing Tropical Forests: Historical Perspectives on Today's Challenges in Asia, Australasia, and Oceania*, ed. John Dargavel, Kay Dixon, and Noel Semple (Canberra: Centre for Resource and Environmental Studies, 1988), 219–29; Richard P. Tucker, "Managing Subsistence Use of the Forest: The Philippines Bureau of Forestry, 1904–60," in *Changing Pacific Forests: Historical Perspectives on the Forest Economy of the Pacific Basin*, ed. John Dargavel and Richard Tucker (Durham, N.C.: Duke University Press, 1992), 105–15; and Bankoff, "One Island Too Many."

64. Florencio Tamesis, "Philippine Forests and Forestry," *Unasylva* 2, no. 6 (1948): 316–25, esp. 320.

65. *Annual Report of the Director of Forestry* (1937), 238–39.

66. Tiburcio Severo, Florencio Asiddao, and Martin Reyes, "Forest Resources Inventory in the Philippines," *Philippine Journal of Forestry* 18, nos. 1–4 (1962): 1–19, esp. 3.

67. On the Japanese invasion of the Philippines, see Teodoro A. Agoncillo, *The Fateful Years: Japan's Adventure in the Philippines, 1941–45*, 2 vols. (Diliman: University of the Philippines Press, 2001). The study was first published in 1965.

68. *Special Study on Timber Resources of the Philippine Islands* (Allied General Headquarters, Southwest Pacific Area, 1944), 2.

69. Hayasi as quoted in Carlos Sulit, "Forestry in the Philippines during the Japanese Occupation," *Philippine Journal of Forestry* 5, no. 1 (1947): 22–47, esp. 24.

70. On Japanese perceptions of Philippine timber resources, see Takayama Keitano, *Nanyo no ringyo* (Tokyo, 1942). I am grateful to Greg Clancey for this reference.

71. Sulit, "Forestry in the Philippines," 29–30. The history of forestry under the Japanese occupation relies almost exclusively on the account of Carlos Sulit, who was a forester at the time. Other short references are evidently based on his report.

72. Ibid., 35.

73. Ibid., 38–39.

74. Agoncillo, *The Fateful Years,* 607–35.

75. Severo et al., "Forest Resources Inventory," 4.

76. The Makiling Forest Reserve (MFR) was established in 1910 and declared a National Botanic Garden in 1920. It is located just south of Manila on the borders between Laguna and Batangas provinces.

77. Sulit, "Forestry in the Philippines," 40–41.

78. Harold A. Winters, with Gerald Galloway, William Reynolds, and David Rhyne, *Battling the Elements: Weather and Terrain in the Conduct of War* (Baltimore: Johns Hopkins University Press, 1998), 1–4.

79. For a discussion of the multifarious ways in which warfare impacts on forests, see J. R. McNeill, "Woods and Warfare in World History," *Environmental History* 9, no. 3 (2004): 388–410; and John Lewallen, *Ecology of Devastation: Indochina* (Baltimore: Penguin Books, 1971).

80. About 75 percent of all U.S. munitions expended during the Korean War (1950–53) was used in this way, and the figure rose as high as 85 percent in Vietnam from 1965 to 1973. Charles H. Southwick, *Global Ecology in Human Perspective* (New York: Oxford University Press, 1996), 315.

81. Michael T. Klare, *Resource Wars: The New Landscape of Global Conflict* (New York: Metropolitan Books, 2001).

82. Thomas F. Homer-Dixon, *Environment, Scarcity, and Violence* (Princeton, NJ: Princeton University Press, 1999), 4; and Robert Kaplan, "The Coming Anarchy," *Atlantic Monthly* 273, no. 2 (1994): 44–76.

83. See Richard P. Tucker and Edmund Russell, *Natural Enemy, Natural Ally: Toward an Environmental History of War* (Corvallis: Oregon State University Press, 2004).

CHAPTER 3

DEVOURING THE LAND

Sherman's 1864–65 Campaigns

Lisa M. Brady

"We have devoured the land," wrote William Tecumseh Sherman in a letter to his wife, Ellen, in June of 1864. "All the people retire before us, and desolation is behind. To realize what war is one should follow our tracks."[1] Sherman was reflecting on the damage wrought by the protracted battle between his Union forces and Confederate Gen. Joe Johnston's army for control over northern Georgia. Neither side had set out to destroy the landscape. The devastation was instead the unavoidable result of armies in motion and one of the inevitable costs of war. Five months later, however, Sherman implemented a strategy, derived from an ancient form of warfare, that shifted devastation from an unintended, haphazard consequence of war into a deliberate and organized weapon of war. Sherman's campaigns through Georgia and the Carolinas from November 1864 to April 1865 focused on destroying the physical, economic, and social landscapes of the region.

Human landscapes are inextricably tied to natural ones. Every society develops unique relationships to the natural world, and the antebellum American South was no exception. As Mart Stewart demonstrated in his study of tidewater Georgia, the plantation system "did not simply organize labor and resources for productive ends . . . but also reorganized nature in the same direction." Plantations, Stewart argued, "constituted agroecological systems that restructured biological processes for agricultural purposes."[2] This in turn created new ecological systems that, beginning in 1861, supported the Confederacy's social, economic, and political structures. In other words, plantation agriculture was the ecological foundation of the Confederacy.[3] This foundation was corrupt, however, based on what Stewart called a "masterful illusion." Although "deeply rooted in experience in the material world," it was predicated on a tenuous power over both slaves and nature.[4] Sherman sought to shatter this "illusion" as he marched his troops across the heart of the Confederacy.

Sherman's 1864–65 campaigns succeeded in revealing the Confederacy's inability to militarily and politically control its own territory by undermining its capacity to marshal critical resources for managing nature in meaningful and productive ways. The marches displayed Union power over the Confederate government, its army, and its residents by denying them the ability to transform nature into culture, thus depriving them of a fundamental source of security.[5] In his Savannah and Carolinas campaigns, Sherman intended to make the fertile Southern landscape into a weapon against the Confederacy by imposing his will on that landscape.

What became one of the most celebrated and condemned campaigns of the war was both a military and psychological triumph. Sherman recognized this, claiming in a letter to Grant that the march would be "a demonstration to the World, foreign and domestic, that we have a power which [Jefferson] Davis cannot resist."[6] The massive foraging raid, or *chevauchée,* clearly illustrated that Southerners' attempts to manipulate the landscape were futile in the face of Federal might. "This may not be war, but rather Statesmanship," Sherman proclaimed, "proof positive that the North can prevail."[7] Sherman's campaigns symbolized and made tangible the authority of the Federal government and confirmed that power over the environment was linked to other kinds of power.

At the core of Sherman's display of force were Special Field Orders No. 120. Issued at Kingston, Georgia, on November 9, 1864, these orders established the protocol for the operation. They divided the Military Division of the Mississippi into two wings, each comprised of two corps that would march along four nearly parallel roads between Atlanta and Savannah.[8] In addition to the necessary pontoon train, each wing could take only a limited number of the wagons reserved for hauling ammunition and providing ambulance service. These organizational details of the campaign enabled Sherman's men to accomplish the main purpose of the march—to demonstrate Federal power to move across and transform the Southern landscape at will. According to Sherman's orders, the Union army would march through the heart of Georgia, taking what it needed, destroying what it didn't, at times acting with magnanimity, at others with callous disregard, all to convince the rebel population that it could not resist the power of the Federal government.

The most important aspect of the campaign revolved around acquiring provisions. Section 4 of the orders explicitly delimited the procedure for collecting rations and forage:

> The army will forage liberally on the country during the march. To this end, each brigade commander will organize a good and sufficient foraging party, under the command of one or more discreet officers, who will gather, near the route traveled, corn or forage of any kind, meat of any kind, vegetables, corn-meal, or whatever is needed by the command, aiming at all times to keep in the wagons at least ten days' provisions for his command, and three days' forage.

> Soldiers must not enter the dwellings of the inhabitants, or commit any trespass; but, during a halt or camp, they may be permitted to gather turnips, potatoes, and other vegetables, and to drive in stock in sight of their camp.

Section 5 restricted to corps commanders the authority to destroy buildings and instructed them to use that authority only when "guerrillas or bushwhackers molest our march, or should the inhabitants . . . manifest local hostility." If such opposition arose, Sherman gave his commanders permission to "order and enforce a devastation more or less relentless, according to the measure of such hostility." Section 6 instructed the cavalry and artillery to appropriate horses, mules, and wagons "freely and without limit," although Sherman urged his men to discriminate between "the rich, who are usually hostile, and the poor and industrious, usually neutral or friendly." Section 7 stated, "Negroes who are able-bodied and can be of service to the several columns may be taken along; but each army commander will bear in mind that the question of supplies is a very important one, and that his first duty is to see to those who bear arms." Section 8 called for the creation of a pioneer battalion for each corps, "composed if possible of negroes [sic]."

Sherman designed each of these provisions to facilitate his march through Georgia and as direct attacks against the state's cultural and physical landscape. The same orders and intentions guided Sherman's operations in South and North Carolina, and attained the same results: the campaigns through Georgia and the Carolinas reasserted Federal power in the region by undermining the local population's ability to manipulate the landscape. In taking away their tools, animals, slaves, and produce, Sherman's forces interrupted—or completely rearranged—Southerners' fundamental relationship with nature. In short, Sherman attacked the ecological foundations of the Confederacy, causing it to crumble.

Sherman's stated goal was to dismantle the infrastructure—industrial, agricultural, and transportational—that supported the Confederate Army. Sherman's men set fire to any building, warehouse, or structure that could be used for military purposes; they pried up the railroad tracks, set the ties ablaze, and melted the rails, twisting them into "Sherman neckties"; cotton stores, too, were burned, with the purpose of undermining the Confederacy's ability to finance its war effort. Fire was one of Sherman's greatest tools, which he used to reduce the Confederacy's military assets literally to ashes.

The obliteration of the Confederacy's physical infrastructure was only part of the battle, however, as a surgeon in the Union Army intimated in a letter home: "It seems now we will hold no interior point between Chattanooga and the Gulf, as all railways, foundries, and other public works will be destroyed before this campaign shall end, and much of the country effectually eaten up and desolated."[9] Sherman believed that he was fighting not only "hostile armies, but a hostile people," and that the war could not be won until both were conquered.[10] His secondary goal, therefore, was to demonstrate the futility of civilian resistance to the Union Army's overwhelming power. He intended to achieve this goal by conducting a *chevauchée,* which would not only destroy much of the region's military and agricultural potential but would also severely damage its ecological foundations. The massive foraging raid would strike at the heart of the residents' relationship with nature, eliminating their power to reorganize the landscape to suit their—or the Confederacy's—needs. Sherman determined to march his army through the heart of Secessia, living off the land, leaving the local residents little except food for thought.

Two days before the great march began, staff officer George Ward Nichols wrote: "I never heard that manna grew on the sand-beaches or in the marshes, though we are sure that we can obtain forage on our way; and I have reason to know that General Sherman is in the highest degree sanguine and cheerful—sure even of success."[11] Sherman had good reason to be optimistic. He was intimately familiar with the physical and cultural geography of the region through which his army would march, having participated in a detailed survey of the area as a young army officer years before.[12] Furthermore, Federal foraging parties reconnoitered around Atlanta in September and October, obtaining new intelligence about central Georgia's circumstances that supported Sherman's decision to cut loose from his supply base.[13]

"They don't know what war means," Sherman wrote to Chief of Staff Henry Wager Halleck, "but when the rich planters of the Oconee and Savannah [rivers] see their fences and corn and hogs and sheep vanish before their eyes they will have something more than a mean opinion of the 'Yanks.' Even now our poor mules laugh at the fine corn-fields, and our soldiers riot on chestnuts, sweet potatoes, pigs, chickens, &c." Sherman concluded that it would "take ten days to finish up our road, during which I will eat out this flank and along down the Coosa, and then will rapidly put into execution the plan."[14] Confident of ample provisions for his men and animals, Sherman set fire to the railroad depot, armory, and cotton warehouses in Atlanta and moved out with his sixty thousand men on November 15, 1864.

Sherman's route took his troops through diverse landscapes ranging from the hilly terrain around Atlanta, through the slopes and ridges of central Georgia, to the marshy lowlands approaching Savannah and the sea (figure 3.1). Thick woods of oak and pine covered much of the territory along the path, but the amount of timber cleared for farming increased dramatically as Union troops moved southeastward from Atlanta. Good roads facilitated the transport of wagons and artillery, and warm dry weather ensured a swift passage over the red clay and sandy soils of central Georgia.[15]

Marching roughly parallel to the Ocmulgee and Oconee rivers, Sherman's army had few natural obstructions to its progress. Maj. Gen. Gustavus Smith of the Georgia Militia recalled that the "face of the country was open, the roads were in good order, the weather was fine and bracing, the crops had been gathered, and were ready for use; in short, a combination of circumstances favored

Devouring the Land 53

Fig. 3.1 Engraving by Alexander Hay Ritchie, dated 1868, depicting Sherman's troops as they destroyed telegraph and railroad lines on their march to the sea in 1864–65. Library of Congress Prints and Photographs Division.

an easy march for Sherman's army."[16] Likewise, Maj. Gen. Jacob Cox remembered that part of the campaign as pleasant, noting that the weather, excepting two days of snow in late November, "had generally been perfect," and that the "camps in the open pine-woods, the bonfires along the railways, the occasional sham-battles at night, with blazing pine-knots for weapons whirling in the darkness, all combined to leave upon the minds of officers and men the impression of a vast holiday frolic."[17]

Despite its relative ease, the campaign had a very serious purpose. In accordance with section 4 of the Special Field Orders, formal foraging parties accompanied each of the army's two wings. Cox noted that "from barn, from granary and smoke-house, and from the kitchen gardens of the plantations, isolated foragers would hasten by converging lines, driving before them the laden mule heaped high with vegetables, smoked bacon, fresh meat, and poultry." As the army reached evening camp, the soldiers "would find this ludicrous but most bountiful supply train waiting for them at every fork of the road, with as much regularity as a railway train running on 'schedule time.'"[18] Georgia provided well for the Union Army. Sherman attributed this to the state "never before having been visited by a hostile army."[19] Others remarked on the region's abundance as well. Henry Hitchcock described the army's surroundings in his diary: "Good farms along the travelled roads, and crops have all been good. We see hardly any cotton—corn almost exclusively instead—*for which we are much obliged.*"[20]

As Hitchcock discovered, by the time Sherman began his march through Georgia, southern agriculture had adapted to accommodate the Confederacy's war needs. Mobilization of the land resulted in a transition, albeit temporary, away from cash crops and toward grains. The riches of Georgia, however, were quickly depleted. Sherman's army of sixty thousand could not tarry, or it would risk the same fate as those left starving in its wake.[21]

The risk became reality as the army neared Savannah. The last seventy-five miles of the Union army's route passed through country "intersected with mirey [sic] swamps and deep morasses."[22] As one of Sherman's soldiers, Theodore Upson, recorded in his diary, "This whole country is a marsh." The enemy had "cut the dykes," Upson observed, and "Evry thing [sic] is a black muck."[23] Sherman's engineers and pioneer corps worked hard to make the sodden landscape traversable, building bridges, corduroying roads, and removing wagons and cannon from the mud.[24]

The city's geography thus favored its defenders. "The strength of Savannah lies in its swamps which can only be crossed by narrow causeways all of which are swept by heavy artillery."[25] A city of nearly twenty-five thousand residents in 1864, Savannah sits on a sandy plateau, approximately forty feet above sea level, between the mouths of the Savannah and Ogeechee rivers. The surrounding land, according to Jacob Cox, "sinks almost to the level of the sea." The coast "is low and cut into islands by deep sinuous natural canals or creeks," which at the time were "widely bordered by the salt marsh which is all awash at high tide." Savannah's location was "almost like an island in the swamps." The only western approach to the city by land consisted of a narrow tongue of land six to eight miles wide between the two rivers, dotted with plantations "in the midst of broad rice-fields which had been reclaimed from the surrounding marsh."[26] These watery fields edged the rivers, creating "a natural barrier around the city on the northwest, about three miles away." The Little Ogeechee River flows between the two larger rivers, providing yet another line of defense southwest of the city. To the north "a series of suburban plantations with their rice-fields in front" made up a "natural line of defence for the town."[27] Surrounded by water—rice fields to the north, west, and south; large rivers to the north and south, and the Atlantic to the east—Savannah was a veritable fortress.

While Sherman's engineers were attempting to transform the flooded landscape into one more suitable for sustaining an army, Confederate General Hardee in Savannah was planning his escape. Not wanting to be trapped inside the city, Hardee evacuated his men on December 19, with only a nominal show of force. The Federal forces moved into the city on the morning of the twenty-first, and Sherman telegraphed President Lincoln, presenting Savannah to him "as a Christmas-gift."[28] Sherman's nearly bloodless campaign—his "demonstration to the world"—had ended in a grand success.

Sherman and his men were well aware of the effects, both military and psychological, that their recent campaign had produced. Henry Hitchcock wrote in his diary that the campaign had proved "that a large army can march with impunity through the heart of the richest rebel state, after boldly cutting loose from all its bases, and subsisting on the country." Hitchcock concluded that the campaign was a "great and important success, full of significance for the future."[29] Another soldier predicted that the campaign "will be one of the really historical campaigns of the war, much more so than some where vastly more fighting was done. It was brilliant in conception and well executed, but practically one of the easiest campaigns we have had."[30]

In less than one month Sherman had marched an army sixty thousand strong through the heart of enemy territory with little resistance. The Military Division of the Mississippi left the foothills of Atlanta on November 15, 1864, traversed the rolling country near Milledgeville, crossed the swampy lowlands to the seacoast ("foraging liberally" along the way), and captured Savannah on December 21. Sherman understood that taking or destroying everything associated with agricultural production would bring the Confederacy to its knees. "I know my Enemy," he wrote to his brother Philemon Ewing, "and think I have made him feel the Effects of war, that he did not expect, and he now Sees how the Power of the United States can reach him in his innermost recesses."[31] The Confederacy's strength as well as its weakness stemmed from its power to transform the environment into a productive landscape, capable of meeting both military and civilian needs. By trumping that power, Sherman's March to the Sea struck a significant blow to the South's agroecological foundation and to the rebellion.

Sherman's men also recognized the consequences of such a strategy. The majority of Union soldiers and officers came from rural communities, and nearly half were farmers or farm laborers; so the destruction of the South's agrarian infrastructure resonated strongly with them.[32] Recounting the damage done and the provisions taken, one officer noted: "We have torn up and destroyed about 200 miles of railroad, burned all bridges and cleaned up the country generally of almost every thing upon which the people could live.... In fact, as we have left the country I do not see how the people can live for the next two years."[33] Others, too, made grim predictions about the region's future. "One thing is for certain," George Nichols wrote, "neither the West nor the East will draw any supplies from the counties in this state traversed by our army for a long time to come. Our work has been the next thing to annihilation."[34]

The damage done to the South's military resources—both provisions and matériel—was extensive. As Sherman told Henry Halleck, "our campaign of the last month, as well as every step I take from this point northward, is as much a direct attack upon Lee's army as though we were operating within the sound of his artillery."[35] He reported that his army had destroyed the Georgia State Railroad entirely and that his soldiers had consumed the corn, sweet potatoes, fodder, cattle, hogs, sheep, and poultry found in the "thirty miles on either side of a line from Atlanta to Savannah." His army requisitioned tens of thousands of horses and mules and ordered thousands more killed in order to prevent

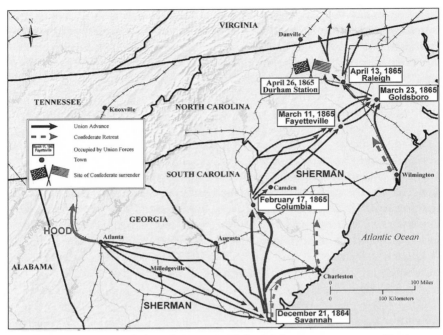

3.1 Map of Sherman's march through Georgia and the Carolinas, 1864–65, created by Aaron Sheehan-Dean. Reproduced by kind permission of Oxford University Press from A Concise Historical Atlas of the U.S. Civil War (2009).

them from falling into enemy hands. A "countless number" of slaves who were determined to seize their freedom joined the moving columns, despite Sherman's exhortations to stay where they were. Sherman estimated the damage at "$100,000,000; at least 20,000,000 of which has inured to our advantage." The rest Sherman chalked up to "simple waste and destruction" (map 3.1).[36]

Daniel Oakey, captain of the 2nd Massachusetts Volunteers, remarked that the Georgia campaign was seen as "a grand military promenade, all novelty and excitement." He believed it had a deeper significance, however, stating that its "moral effect on friend and foe was immense. It proved our ability to lay open the heart of the Confederacy, and left the question of what we might do next a matter of doubt and terror."[37] Emma LeConte, a young resident of Columbia, South Carolina, certainly wondered what lay in store for her native state. "Georgia has been desolated," she wrote in her diary on New Year's Eve, 1864. "The resistless flood has swept through that state, leaving but a desert to mark its track." LeConte had heard that Sherman's men were "preparing to hurl destruction upon the State they hate most of all, and Sherman the brute avows his intention of converting South Carolina into a wilderness."[38] Indeed, exhilarated by his success in Georgia, Sherman wrote to Grant in late December exclaiming, "I could go on and smash South Carolina all to pieces."[39]

Devouring the Land

Grant's initial plan was for Sherman to move north from Savannah by water to join him in Virginia against Lee, but the success of the Georgia campaign and the lack of ready transports persuaded Grant that a foray through the Carolinas, "if successful," would promise "every advantage."[40] With both Georgia and the Carolinas unable to provide supplies, Lee's forces at Richmond would have to rely on a small area of Virginia to feed and support them. Noting that "although [Virginia] was fertile, it was already well exhausted of both forage and food," Grant "approved Sherman's suggestion . . . at once."[41]

Sherman's army left Savannah on January 19, 1865, and arrived on the coast of South Carolina four days later. Daniel Oakey described the impending campaign as "formidable," one that would involve "exposure and indefatigable exertion."[42] Sherman's plan for the Carolinas mirrored his actions in Georgia, with four corps organized into two wings marching north along nearly parallel lines from the coast toward Columbia. From there he would march to Goldsboro, North Carolina, and then to Raleigh, where he would end his "demonstration." Oakey noted that the campaign's success depended on continuous forward movement, "for even the most productive regions would soon be exhausted by our 60,000 men and more, and 13,000 animals." He further remarked that despite being fully prepared for "a pitched battle, our mission was not to fight, but to consume and destroy."[43] Special Field Orders No. 120 remained in effect for the entire campaign, including the injunction to implement them with restraint and order.

The march through Georgia took Sherman's army across rolling, fertile countryside; that through the Carolinas, South Carolina especially, covered less productive and more unforgiving terrain. In addition to being a more hostile landscape, South Carolina was known for its secessionist fervor. Pro-Union sentiment in Georgia, while perhaps not widespread, exceeded that in South Carolina, and Sherman and his soldiers reserved most of their vehemence for the hotbed of rebellion.[44] South Carolina, in contrast to Georgia, struck back. Resembling the area around Savannah, where rice fields mingled with swamps and streams, and rivulets meandered across the landscape creating islands out of higher ground, coastal South Carolina was wet and marshy with few good roads. Maj. Henry Hitchcock described them as "frequently bad," cutting through sparsely settled and "very uninviting country." They were "unusually closely bordered either by the dense woods full of almost impenetrable underbrush, or marshes, swamps and wet rice-fields, on both sides."[45]

Heavy rains throughout January complicated things further, keeping the Union troops practically immobilized. The left wing included the XIV and XX Corps and was delayed forty miles north of Savannah at Sister's Ferry, where it grappled with the flooded Savannah River. The XV and XVII Corps comprised the right wing, which was stranded at Beaufort, South Carolina, unable to move its supply trains and artillery over the flooded terrain. Sherman's troops desperately needed solid ground, but the continued rains caused rivers and streams to swell beyond their banks, some spreading a mile or more.

From the beginning of the campaign, then, water—not rebel forces—posed the greatest challenge to Sherman's army. Describing the problems faced by the left wing at Sister's Ferry, Jacob Cox noted that the "almost continuous rains" threatened the trestle and pontoon bridges, and that the Union causeway out of Savannah "was under water, and the whole region was more like a great lake than a habitable land."[46]

After nearly a week at Beaufort, Sherman pressed northward, only to be stopped again by rain and impassable muddy roads at Pocotaligo. On January 29, Sherman reported to Grant: "terrific Rains ... caught me in motion, and nearly drowned some of my Columns in the Rice fields of the savannah, swept away our Causeway which had been carefully corduroyed, and made the Swamps here about mere lakes of slimy mud."[47] Situated midway between the Coosawatchie and Salkahatchie rivers approximately ten miles in from the South Carolina coast, Pocotaligo was a small community where Union troops had captured a fort earlier that month. The final days of January brought a reprieve from the constant precipitation, so Sherman ordered his troops to move out on February 1, believing he was on "terra firma."[48]

Sherman discovered, though, that appearances could be deceiving and that even seemingly solid ground posed dangers. The route from Pocotaligo to Columbia was "practically determined by the topography of the country, which, like all the Southern seaboard, is low and sandy, with numerous extensive swamps and deep rivers widely swamp-bordered, only approachable by long causeways."[49] These "made roads," as Henry Hitchcock described them, frequently required extensive corduroying to make them usable.[50] The very nature of the ground did more to impede the progress of Sherman's troops than any Confederate force.

Rain was a significant factor throughout most of the campaign, hampering forward progress. Jacob Cox recalled the exertion required to drag loaded wagons and heavy artillery "over mud roads in such a country" and "the infinite labor required to pave these roads with logs, levelling the surface with smaller poles in the hollows between, adding to the structure as the mass sinks in the ooze." Multiply that work by the hundreds of miles the army had to cover, and one would "get a constantly growing idea of the work, and a steadily increasing wonder that it was done at all," Cox wrote.[51]

The purpose of Sherman's campaign through the Carolinas was not, however, to pave the area's roads; rather, the difficult labor was a prerequisite to the successful demonstration of the Federal Army's power. As in Georgia, Sherman's primary goal was to disrupt the Confederacy's supply lines. The most effective way to accomplish that goal was to employ the same strategy he had used in Georgia: Sherman attacked the states' ecological foundations.

Special Field Orders No. 120 again provided the bounds for Sherman's strategy and enabled him to deliver a mortal blow to the Confederacy. Early in the campaign Lt. Gen. Wade Hampton, lately placed in charge of defending his home state of South Carolina, condemned Union foraging as a violation of the rules of war. "Of course you cannot question my right to 'forage on the country,'" Sherman responded. "It is a war right as old as history. The manner of exercising it varies with circumstances, and if the civil authorities will supply my Requisitions, I will forbid all foraging." However, Sherman concluded, "I find no civil authorities who can respond to calls for Forage or provisions, and therefore must collect directly of the People."[52] Sherman's references to law and history belied his intention to make Southerners "feel the hard hand of war."[53]

Sherman's men shared his convictions. One argued that the Union's task "was incomplete while the Carolinas, except at a few points on the sea-coast, had not felt the rough contact of war." He voiced concern, though, that "their swamps and rivers, swollen and spread into lakes by winter floods, presented obstructions almost impracticable to an invading army, if opposed by even a very inferior force."[54] George Nichols implicitly equated the hurdles nature presented with the challenge of fighting and defeating the rebel forces themselves. He recorded in his journal the "remarkable experience" of "floundering" through South Carolina's swamps. Surrounded by water and sodden ground, "our tireless soldiers stop for nothing," Nichols wrote. "Yesterday afternoon the swamps were conquered, the Salkahatchie was crossed, and a force of the enemy who offered a determined opposition to our passage of the stream were driven back."[55]

That the threat posed by rebel forces seems an afterthought to Nichols is telling, but not surprising. As Sherman's army moved north through South Carolina toward Columbia, the greatest challenges it faced came not from armed men but from an inhospitable and generally uncooperative environment. In a rare moment of lyrical hyperbole, Jacob Cox summed up his impression of the South Carolina campaign:

> If the march through Georgia remained pictured in the soldiers' memories as a bright, frolicsome raid, that through South Carolina was even more indelibly printed as a stubborn wrestle with the elements, in which the murky and dripping skies were so mingled with the earth and water below as to make the whole a fit type of "chaos come again"; but where, also, the indomitable will of sixty thousand men, concentrated to do the inflexible purpose of one, bridged this chaos for hundreds of miles, and out-laboring Hercules, won a physical triumph that must always remain a marvel.[56]

In an era when man's right and ability to control the natural environment were assumed, the obstacles of nature must have been great indeed to evoke the image of Hercules. In the end Sherman's men were not simply marching over South Carolina's territory; they were fighting it all the way.

According to historian Russell Weigley, South Carolina's "best bet against [Sherman] seemed to be not a scattered army and either worn-out or inexperienced troops, but geography and weather."[57] Both seemed prepared to aid the Confederates in blocking Sherman's advance. Rain fell on twenty-eight of the first forty-five days of the campaign (which lasted only fifty-five days in total from Beaufort to Fayetteville, North Carolina), complicating the numerous stream crossings and exacerbating the already difficult task of gathering rations. The swampy landscape precluded agricultural pursuits in much of the region, and even those "acres in the armies' path as happened to be dry were not much more hospitable to farming than the marshes."[58] Foragers thus had to be aggressive and creative in performing their duties during the campaign through South Carolina.

Disparities in production levels in the areas between Sherman's marching columns mattered little to Union foragers. They took everything they found—and they found everything. Attuned to any disturbance of the soil, and wise to common hiding places, foragers literally left no stone unturned. "The fresh earth recently thrown up, a bed of flowers just set out, the slightest indication of a change in appearance or position, all attracted the gaze of these military agriculturalists. It was all fair spoil of war, and the search made one of the excitements of the march."[59] The foragers—or bummers, as they came to be known—made no distinction between rich or poor, black or white. Their determination to wreak havoc on South Carolina increased as they neared its capital city, Columbia. The foragers' actions (and reputations) prompted Columbia resident Joseph LeConte to write, "The enemy, swearing vengeance against South Carolina, the cradle of Secession, is approaching step by step, consternation and panic [sic] flight of women and children in front, and a blackened ruin behind."[60]

On February 17, 1865, the four separate columns of Sherman's army converged and entered Columbia. Three days later Sherman moved his four columns out of the capital city toward North Carolina, where he would end his display of power against the Confederacy. Again, adverse weather conditions and unreliable sources of forage presented the biggest challenges for the Union troops. Rain continued to fall, turning the roads into quagmires, and spotty food supplies left the men hungry. The ferocity with which Union foragers collected provisions in South Carolina diminished once they crossed the border into North Carolina. The ends were the same, however; Sherman's troops scoured clean the Carolinas, leaving the landscape "in no condition to supply another army which may need to pass over it, if indeed it can supply the necessaries of life for the people living in it."[61]

Sherman's army was not the only one on North Carolina territory, however. After fifty-five days of slogging through mud, swamps, and rain, Sherman once again faced the formidable Joe Johnston. For the first time since their bloody confrontations in northern Georgia, Sherman's and Johnston's armies clashed at Averasboro, North Carolina, on March 16 and again at Bentonville three days later.[62] The Confederate Army sustained heavy casualties, leaving Johnston only twenty-one thousand troops to defend Goldsboro against Sherman's sixty thousand hardened veterans. In the face of such odds, Johnston chose to

Devouring the Land

retreat; Sherman captured Goldsboro then moved on toward Raleigh. In Weigley's words, it seemed that "against any possible human intervention, Sherman's armies were irresistible."[63]

Sherman's troops may have appeared invincible, but their resilience to nature's onslaught diminished as the campaign dragged on. Jacob Cox recalled that the days following the battle of Averasboro "are remembered by the officers and men ... as among the most wearisome of the campaign. Incessant rain, deep mud, roads always wretched but now nearly impassable, seemed to cap the climax of tedious, laborious marching."[64] It would also be the last major battle Sherman's troops would fight. On April 26, 1865, two weeks after Grant accepted Lee's surrender of the Army of Northern Virginia, Sherman accepted Johnston's surrender at Durham Station.

Sherman's campaigns through Georgia and the Carolinas left behind an awesome spectacle of destruction. For some of Sherman's men, like Daniel Oakey, these scenes verged on the sublime. Describing the army's advance into "the wild regions of North Carolina," he wrote:

> The scene before us was very striking; the resin pits were on fire, and great columns of black smoke rose high into the air, spreading and mingling together in gray clouds, and suggesting the roof and pillars of a vast temple. All traces of habitation were left behind, as we marched into that grand forest with its beautiful carpet of pine-needles. The straight trunks of the pine-tree shot up to a great height, and then spread out into a green roof, which kept us in perpetual shade. As night came on, we found that the resinous sap in the cavities cut in the trees to receive it, had been lighted by "bummers" in our advance. The effect of these peculiar watch-fires on every side, several feet above the ground, with flames licking their way up the tall trunks, was peculiarly striking and beautiful.[65]

Despite the scene's allure, however, Oakey concluded that the "wanton" destruction was "sad to see," all the more so because the "country was necessarily left to take care of itself, and became a 'howling waste.'"[66]

Rural or urban, wild or domestic—no landscape in his path was immune from Sherman's destructiveness. The war damage to both the built and natural landscapes of the South was obvious, and sources describing the destructiveness of Sherman's *chevauchées* abound. Sherman himself recalled one house near Pocotaligo, South Carolina, "with a majestic avenue of live-oaks, whose limbs had been cut away by the troops for firewood."[67] Observing a similar scene, Henry Hitchcock sardonically remarked, "It is bad for the live oaks and cedars that so many soldiers are camped round here in cold weather."[68] Used for other purposes as well, trees were frequent targets of the soldier's ax. Confederate soldiers felled trees across the roads in South Carolina to slow Union progress, and Union troops cut them down to speed their marches through the swamps. In one instance trees "quickly became logs" to rebuild a bridge over a flooded river. "No matter if logs disappeared in the floating mud; thousands more were coming from all sides," placed "layer by layer" to build a "wooden causeway" over the aptly named Lumber River.[69]

The primary target, however, of the March to the Sea and the campaigns through the Carolinas, was the root of Confederate power—the rich agricultural lands of the Deep South. To attack this resource, Sherman refined the ancient *chevauchée* and implemented it on a grand scale. His foragers carried out sections 4 through 7 of Special Field Orders No. 120 with awesome zeal, leaving local citizens struggling to feed themselves and with bleak prospects ahead. Carving a path sixty miles wide in Georgia and nearly as wide in the Carolinas, Sherman's final campaigns were instrumental in the drive to defeat the Confederacy. The swath of devastation was to the war as a back burn is to a wildfire: isolating the valuable resources of Georgia and the Carolinas from the rest of the Confederacy prevented these states from feeding the fire of rebellion.

The campaigns successfully destroyed the last stores available to the dwindling rebel armies and sent a powerful message to the Confederate populace about the reach of Federal power. They demonstrated that the Federal government ultimately controlled how the American landscape and its resources would be used and by whom. The strategy was at once subtle and overt, and its message was unmistakable. Sherman's *chevauchées* through the Deep South illustrated the Union's power to determine the face of the American landscape and delivered a fatal blow to the rebellion.

The nature of Sherman's strategy—to "forage liberally off the land"—targeted the South's agricultural landscape, its most important resource and the basis of its economy, society, and identity. Union troops destroyed or confiscated cotton, food, forage, crops, livestock, and farming implements. The immediate effects of these campaigns were obvious. Alexander Lawson, a Confederate prisoner for part of the march through Georgia, recalled that Sherman had "made brags that he would make a black mark to the sea. He certainly did." Lawson escaped just outside of Savannah and turned back on Sherman's path. "I found nothing," Lawson wrote, "no hogs, cattle, sheep, chicken, or anything else to eat. I saw a number of the very finest ladies in Georgia in the camps picking up grains of corn for the purpose of sustaining life, who a week before that did not know what it was to want for anything. I finally crossed the Savannah River into South Carolina, where his army hadn't been, and it was the first food that I had for about eighteen days." For these depredations, Lawson believed that Sherman had earned a "warm spot in Hell."[70]

Civilians, too, recorded their reactions to the ruin left in Sherman's wake. Eliza Andrews "almost felt as though [she] should like to hang a Yankee" after witnessing the damage done to her native state. Near Sparta, Georgia, twenty-five miles northeast of Milledgeville, she came upon the "burnt country": "There was hardly a fence left standing all the way from Sparta to Gordon. The fields were trampled down and the road was lined with carcasses of horses, hogs, and cattle that the invaders, unable either to consume or to carry away with them, had wantonly shot down, to starve out the people and prevent them from making their crops." Calling the passing troops "savages," Andrews excoriated them for leaving no grain "except [the] little patches they had spilled when feeding their horses." Remarking that "there was not even a chicken left in the country to eat," Andrews claimed that a "bag of oats might have lain anywhere along the road without danger from the beasts of the field, though I cannot say it would have been safe from the assaults of a hungry man."[71] Farther into her journey, Andrews observed a field near Milledgeville that had been used as a camp by thirty thousand Union soldiers, "strewn with the debris they had left behind." Andrews noted that "the poor people of the neighborhood were wandering over it, seeking for anything they could find to eat, even picking up grains of corn that were scattered around where the Yankees had fed their horses." In the midst of the wreckage, however, Andrews saw some men "plowing in one part of the field, making ready for next year's crop."[72]

As Andrews's account reveals, for most of the South, the destruction of crops elicited only temporary consequences. Sherman did not literally destroy the land. He did not salt the earth, as one officer suggested he do. Nor did his men poison wells, as they were accused of doing.[73] What was laid waste, however, was the ecological foundation of the Confederacy. Slave-based agriculture served as the cornerstone of the South's relationship to the natural environment. The mortar that held it in place—an insecure system of control grounded in the oppression of black Americans and on constant and delicate negotiations with nature—could not hold in the face of a direct attack. Sherman's *chevauchées* through Georgia and the Carolinas capitalized on the tenuous character of the Southern agricultural system, shifting the balance of power just enough to cause the Confederacy to topple in upon itself.

Notes

1. Letter, William T. Sherman to Ellen Ewing Sherman, June 26, 1864, quoted here from *Sherman's Civil War: Selected Correspondence of William T. Sherman, 1860–1865*, ed. Brooks D. Simpson and Jean V. Berlin (Chapel Hill: University of North Carolina Press, 1999), 657.

2. Mart Stewart, *"What Nature Suffers to Groe": Life, Labor, and Landscape on the Georgia Coast, 1680–1920* (Athens: University of Georgia Press, 1996), 90.

3. By the term *ecological foundation* I mean the interrelated set of systems—economic, political, and social—that shape and are in turn shaped by human relationships with the natural world. For example, economic and ecological systems are inextricably linked, an argument clearly illustrated in William Cronon's study of colonial New England, *Changes in the Land: Indians, Colonists, and the Ecology of New England* (New York: Hill and Wang, 1983). Donald Worster has made the compelling argument that modes of production "have been engaged not merely in organizing human labor and machinery but also in transforming nature" ("Transformations of the Earth," *Journal of American History* 76 [March 1990]: 1087–1106, esp. 1090).

4. Stewart, "What Nature Suffers to Groe," 191–92.

5. This was not the first attempt by Union generals to demonstrate this power. Ulysses S. Grant inadvertently discovered the utility of the strategy when, in 1862, Confederate General Van Dorn destroyed Grant's supply depot at Holly Springs, Mississippi, forcing the Union army to live off the land. Grant implemented the same strategy the following year when he moved his troops across the Mississippi River to lay siege to Vicksburg.

6. Letter, W. T. Sherman to Ulysses S. Grant, November 6, 1864, quoted here from Simpson and Berlin, eds., *Sherman's Civil War*, 751.

7. Ibid.

8. See "Special Field Orders, No. 120" in William T. Sherman, *Memoirs of General W. T. Sherman*, 2 vols. (1875; rpt. New York: Library of America, 1990), 2:651–53. All subsequent quotations of Sherman's *Memoirs* follow this edition.

9. This letter, attributed only to "a Chicago surgeon in General Sherman's Army," was published in the December 2, 1864, issue of the [*Natchez*] *Weekly Courier* (Newspapers and Periodicals Division, Library of Congress).

10. Letter, W. T. Sherman to Henry Wager Halleck, December 25, 1864, quoted here from Sherman, *Memoirs*, 705.

11. George Ward Nichols, *The Story of the Great March* (New York: Harper and Brothers, 1865), 37.

12. "Thus by a mere accident I was enabled to traverse on horseback the very ground where in after-years I had to conduct vast armies and fight great battles. That the knowledge thus acquired was of infinite use to me, and consequently to the Government, I have always felt and state" (Sherman, *Memoirs*, 29–30, esp. 30).

13. The *New York Herald* reported that Sherman found that Georgia's "endless cottonfields of 1860 had become her inviting corn fields of 1864, for the subsistence of rebel armies" (*New York Herald*, November 28, 1864, reprinted in the [*Natchez*] *Weekly Courier*, December 9, 1864. Newspapers and Periodicals Division, Library of Congress).

14. Letter, W. T. Sherman to Halleck, October 19, 1864, quoted here from Simpson and Berlin, eds., *Sherman's Civil War*, 736.

15. Maj. Henry Hitchcock kept a detailed journal on the march from Atlanta to Savannah, recording weather, soil composition, terrain, vegetation, and agricultural improvements; the majority of this description is derived from several of Hitchcock's journal entries. See Henry Hitchcock, *Marching with Sherman*, ed. M. A. DeWolfe Howe (New Haven, Conn.: Yale University Press, 1927), 60, 65, 76, and 90. See also Jacob D. Cox, *Sherman's March to the Sea: Hood's Tennessee Campaign and the Carolina Campaigns of 1865*, vol. 12 of *Campaigns of the Civil War*, 13 vols. (1881–83; rpt. New York: Da Capo Press, 1994), 25–26.

16. Gustavus W. Smith, "The Georgia Militia during Sherman's March to the Sea," in *The Way to Appomattox*, vol. 4 of *Battles and Leaders of the Civil War*, ed. Robert Underwood Johnson and Clarence Clough Buel, abridged ed. (New York: Appleton-Century-Crofts, 1956), 667. Smith was a major general in the Georgia militia.

17. Cox, *Sherman's March to the Sea*, 42.

18. Ibid., 39, 40.

19. Sherman, *Memoirs*, 658.

20. Hitchcock, *Marching with Sherman*, 108 (journal entry for November 28, 1864).

21. Cox, *Sherman's March to the Sea*, 22.

Devouring the Land 65

22. George Cram to Mother, December 18, 1864, in *Soldiering with Sherman: Civil War Letters of George F. Cram,* ed. Jennifer Cain Bohrnstedt (DeKalb: Northern Illinois University Press, 2000), 151.

23. Theodore Frelinghuysen Upson, *With Sherman to the Sea: The Civil War Letters, Diaries & Reminiscences of Theodore F. Upson,* ed. Oscar Osburn Winther (Bloomington: Indiana University Press, 1958), 139 (diary entry for December 11, 1864).

24. Cox, *Sherman's March to the Sea,* 41–42.

25. Letter, W. T. Sherman to Ellen Ewing Sherman, December 16, 1864, quoted here from Simpson and Berlin, eds., *Sherman's Civil War,* 768.

26. Cox, *Sherman's March to the Sea,* 43.

27. Ibid., 43–44.

28. Sherman, *Memoirs,* 711.

29. Hitchcock, *Marching with Sherman,* 167 (journal entry for December 10, 1864).

30. Letter, Thomas Osborn to S. C. Osborn, December 31, 1864, quoted here from Thomas Ward Osborn, *The Fiery Trail: A Union Officer's Account of Sherman's Last Campaigns,* ed. Richard Harwell and Philip N. Racine (Knoxville: University of Tennessee Press, 1986), 80.

31. W. T. Sherman to Philemon B. Ewing, January 29, 1865, quoted here from Simpson and Berlin, eds., *Sherman's Civil War,* 810.

32. See Phillip Shaw Paludan, "Agriculture and the Benefits of War," in *A People's Contest: The Union and Civil War, 1861–1865,* 2nd ed. (Lawrence: University Press of Kansas, 1996), 151–69.

33. Thomas Osborn to Abraham Osborn, December 14, 1864, in Harwell and Racine, eds., *Fiery Trail,* 47.

34. Nichols, *Story of the Great March,* 81 (journal entry for December 3, 1864). See also Joseph T. Glatthaar, "Foraging," in *The March to the Sea and Beyond: Sherman's Troops in the Savannah and Carolinas Campaigns* (Baton Rouge: Louisiana State University Press, 1985), 119–33; and Mark Grimsley, "The Limits of Hard War," in *The Hard Hand of War: Union Military Policy toward Southern Civilians, 1861–1865* (Cambridge: Cambridge University Press, 1995), 171–204.

35. W. T. Sherman to Halleck, December 24, 1864, quoted here from Sherman, *Memoirs,* 704.

36. W. T. Sherman to Halleck, January 1, 1865, quoted here from *The War of the Rebellion: A Compilation of the Official Records of the Union and Confederate Armies,* 70 vols. in 128 (Washington, DC: Government Printing Office, 1880–1901), 44:7–15, esp. 14 (hereafter cited as *OR*). The modern dollar equivalent would be approximately eleven times this amount, according to the Economic History Resources Web site, which calculates relative historic values of the American dollar; see Economic History Resources, "What Is Its Relative Value in U.S. Dollars," at http://www.eh.net/hmit/compare/, accessed November 23, 2004.

37. Daniel Oakey, "Marching through Georgia and the Carolinas," in Johnson and Buel, eds., *Battles and Leaders,* 4:671–79, esp. 674.

38. Emma LeConte, *When the World Ended: The Diary of Emma LeConte,* ed. Earl Schenck Miers (New York: Oxford University Press, 1957), 3–4 (diary entry for December 31, 1864, Columbia, S.C.).

39. W. T. Sherman to Grant, December 22, 1864, quoted here from Simpson and Berlin, eds., *Sherman's Civil War,* 722. Sherman refers to Thomas's success at Nashville.

40. Ulysses S. Grant, *Memoirs and Selected Letters: Personal Memoirs of U.S. Grant, Selected Letters, 1839–1865*, 2 vols. in 1 (New York: Library of America, 1990), 671.

41. Grant, *Memoirs*, 672.

42. Oakey, in Johnson and Buel, eds., *Battles and Leaders*, 4:675.

43. Ibid., 4:674–75.

44. For an excellent examination of both the subtle and obvious differences in Union soldiers' implementation of Special Field Orders No. 120 as they applied to Georgia and South and North Carolina, see Grimsley, *Hard Hand of War*, 171–204.

45. Hitchcock, *Marching with Sherman*, 229 (journal entry for January 30, 1865).

46. Cox, *Sherman's March to the Sea*, 168.

47. W. T. Sherman to Grant, January 29, 1865, Pocotaligo, SC, quoted here from Simpson and Berlin, eds., *Sherman's Civil War*, 817.

48. Ibid.

49. Cox, *Sherman's March to the Sea*, 171–72.

50. Hitchcock, *Marching with Sherman*, 259 (journal entry for February 6, 1865). Hitchcock referred to these roads as "made" because earth and other materials were piled up to provide a ridge of sorts—much like the levees of Louisiana.

51. Cox, *Sherman's March to the Sea*, 171–72.

52. Letter, W. T. Sherman to Lt. Gen. Wade Hampton (CSA), February 24, 1865, quoted here from Simpson and Berlin, eds., *Sherman's Civil War*, 820.

53. Letter, W. T. Sherman to Halleck, December 24, 1864, quoted here from Sherman, *Memoirs*, 705. Although this passage is taken from a letter written just after the Georgia campaign, it illustrates Sherman's rationale for the Carolina campaigns as well.

54. Oakey, in Johnson and Buel, eds., *Battles and Leaders*, 4:674.

55. Nichols, *Story of the Great March*, 136 (entry for February 4, 1865).

56. Cox, *Sherman's March to the Sea*, 172.

57. Russell F. Weigley, *A Great Civil War: A Military and Political History, 1861–1865* (Bloomington: Indiana University Press, 2000), 418.

58. Ibid., 419.

59. Nichols, *Story of the Great March*, 115.

60. Joseph LeConte, *'Ware Sherman: A Journal of Three Months' Personal Experience in the Last Days of the Confederacy* (Berkeley: University of California Press, 1937), 81 (entry for February 8–15 [1865]).

61. Letter, Thomas Osborn to Abraham Osborn, Fayetteville, NC, March 12, 1865, in Harwell and Racine, eds., *Fiery Trail*, 177. For statistics of what Sherman's army took from Georgia, South Carolina, and North Carolina, see Glatthaar, *March to the Sea and Beyond*, 130.

62. Averasboro and Bentonville were located approximately thirty and twenty-seven miles west of Goldsboro, respectively. For a brief description of the fight, see Weigley, *Great Civil War*, 421.

63. Weigley, *Great Civil War*, 422.

64. Cox, *Sherman's March to the Sea*, 184.

65. Oakey, in Johnson and Buel, eds., *Battles and Leaders*, 4:677–78. The process of removing resin from pine trees for rosin and turpentine production involves "tapping," or "bleeding," the trees by cutting V-shaped notches in the xylem (wood stem) to direct the sticky, highly flammable substance into pits cut into the tree for collection.

66. Oakey, in Johnson and Buel, eds., *Battles and Leaders*, 4:678.

67. Sherman, *Memoirs*, 736.

68. Letter, Henry Hitchcock to Mary C. Hitchcock, January 29, 1865, in Hitchcock, *Marching with Sherman*, 219.

69. Oakey, in Johnson and Buel, eds., *Battles and Leaders*, 4:677.

70. Letter, Alex Lawson to Mr. S. A. Cunningham, December 12, 1910, in the collections of the Filson Historical Society, Louisville, Kentucky.

71. Eliza Andrews, "Eliza Andrews Comes Home through the Burnt Country," in *The Blue and the Gray*, ed. Henry Steele Commager, 2 vols. in 1 (New York: Wings Books, 1991), 958–59.

72. Ibid.

73. Chief of Staff Henry Wager Halleck told Sherman as he began his campaign: "Should you capture Charleston, I hope that by *some accident* the place may be destroyed, and if a little salt should be sown upon its site, it may prevent the growth of future crops of nullification and secession" (letter, Halleck to W. T. Sherman, December 18, 1864, in Sherman, *Memoirs*, 700).

CHAPTER 4

ENVIRONMENTS OF DEATH

Trench Warfare on the Western Front, 1914–18

Dorothee Brantz

> The face of the landscape was dark and incredible; the battle had wiped away what was appealing in the area and engraved it with horrible features, from which the solitary observer drew back in fright.
>
> —Ernst Jünger

War is an outdoor activity. Whereas battle strategies might be devised indoors, the actual fighting—to the extent that it involves the conquest of territory—has to take place outside on a battlefield. In order to conquer land, soldiers have to cover physical space. Consequently, every war has a distinct spatiality depending on the terrain and climate of the area and the type of warfare conducted.

Battlefields are not artificially created places but sites of transformation, where a peaceful landscape is gradually turned into an environment of war.[1] I purposefully use the terms *landscape* and *environment* to distinguish between two different notions of space. The term *landscape* has undergone a fundamental redefinition from the early modern legal and territorial conception of *Landschaft* as property to the more modern understanding of landscape as an object of visual representation.[2] A landscape is a more or less natural terrain (that can include forests, fields, towns, cities, mountains) inscribed with a range of cultural relations, not least because a landscape requires an observer who reflects on the scenery or actively creates it through the cultivation of land.[3] In its modern connotation, a landscape is primarily an aesthetic space consisting of a vista and a horizon that is meant to be "*seen*, either framed within a sketch or painting, composed within borders of a map, or viewed from a physical eminence through receding planes of perspective."[4]

An environment, in contrast, is more than visual. One might *look at* a landscape, but one *lives in* an environment. As the original meaning of the

Middle English verb *environ* (to surround) indicates, an environment consists of physical structures and phenomena, both natural and man-made, as well as the atmospheric conditions that envelop every space. All living organisms need a specific environment to sustain their existence. Environments are lived-in areas that demand a more immediate comprehensive engagement of the senses—sight as much as sound, smell, and touch.

While a landscape, as Denis Cosgrove has argued, "acts to 'naturalize' what is deeply cultural," environments serve to acculturate nature.[5] Environments are also permeated by culture, but the more important point is that environments are characterized by the direct and continuous impact of nonhuman physical forces on people's experience. People can live without landscape, but they cannot survive without an environment. Of course, these two terms are conceptual constructs reliant on artificial distinctions between things that cannot really be separated. A landscape always consists of an environment, and most environments have a landscape.[6] They are two dimensions of the same space; nevertheless, this distinction strikes me as essential because it illuminates the changing relationship between humans and the physical environment during times of war when landscapes are turned into battlefields and civilians into soldiers.

The first characteristic of an environment of war is that it has ceased to be merely a landscape through which soldiers move but has turned into a space where soldiers engage in combat. To the extent that war revolves around the conquest of territory, it can be accomplished only through the active engagement with the environment, which raises the questions of how a peaceful landscape becomes an "environment of war" and just what is meant by that phrase. Looking at the example of trench warfare on the western front during World War I, this chapter investigates how this spatial transformation took place and how environmental factors shaped the daily practice of warfare and soldiers' perceptions of the front.[7]

Looking at soldiers' diaries, letters, and memoirs, one quickly realizes that environmental forces played a decisive role in the everyday conduct of war. After all, soldiers on the front lines were exposed not only to enemy armies but also to natural hazards. Whereas weapons killed, the environmental conditions made combat all the more miserable. War was still deadly in sunshine but became even less tolerable during prolonged times of rain. Expressing precisely this sentiment, one German soldier wrote: "For the first time in ten days we have lasting sunshine. That makes everything bearable, but the first weeks were horrible because it rained continuously and everything turned into a swamp. One could not take ten steps without sinking knee-deep into it."[8]

The physical conditions in the trenches were not merely unpleasant but posed real threats to soldiers' lives, particularly during World War I, when the destructive potential of new military technologies began to threaten vast landscapes and natural resources on which entire regions depended. Historians generally agree that World War I initiated a new era of warfare, but they tend to focus on the impact of new technologies and how they led to a new synergy between men and machines; however, I would argue that it was not simply the nexus of men and technology but rather the convergence of men, technology, and the environment that unleashed the unprecedented deadly force of modern mass warfare.

Interestingly, whereas soldiers' accounts are full of references to their surroundings, most of the secondary literature about World War I fails to consider the role of the environment in trench warfare. This scholarly neglect is twofold. Environmental historians have paid surprisingly little attention to warfare.[9] One has only to look at the extensive debates about environmental history or the growing number of introductory texts, most of which do not mention the phenomenon of war at all. In a similar vein, those who study war usually have little to say about the environment. Military historians, and especially scholars of the new military history, have examined a wide array of topics related to warfare, including its strategic, economic, technological, cultural, social, and gender dimensions; few, however, have studied the environmental aspects of war. This is not to say that the natural world is altogether ignored in the histories of World War I. Studies of everyday life in the trenches often describe environmental conditions in great detail; but because they focus primarily on how trench warfare affects social relations—including its impact on the identities of individual soldiers as well as the collective mentality of those who served at the front—these studies treat the environment as merely the backdrop against which the drama of physical and psychological attrition unfolded.[10] Histories of the creative imagination as well as of the memory of war also frequently discuss the role of landscape and space on the western front; however, most of them fail to distinguish between environment, nature, and landscape—in fact they often use the terms interchangeably.[11]

My essay seeks to offer a slightly different reading of soldiers' front-line accounts by paying closer attention to the dual relationship between, on the one hand, the place of the environment in soldiers' everyday experience in the field and, on the other hand, the human impact on their natural surroundings. The chapter analyzes the everyday reality of warfare and how abstract notions about going to war in order to defend a nation turned into the dirty reality of combat. This reality had little to do with heroic landscapes and battles but was characterized by an expanding environment of death and destruction.

Going to War

In August 1914, more than nine million men from every social background, many of them volunteers, were mobilized to form the armies of Europe.[12] In addition to men, millions of animals—mainly horses but also dogs and, for the first time, carrier pigeons—were sent to the front.[13] When the first trains moved toward the front lines, summer was still in full bloom. It is no accident that wars usually break out in spring or summer. Military strategists through the ages have been well aware that favorable weather conditions, apart from easing the movement of troops and matériel, aid the planning of battles. Not only are environmental factors less likely to affect the outcome of the battle, but warm weather also enhances the willingness of soldiers and nations to mobilize. The initial *Kriegsbegeisterung* (enthusiasm for war) that took hold of many nations in 1914 was in part due to the widespread conviction that the war would be over "by the time the leaves fall."[14] In other words, the war would not just be short; it would be fought in fair weather.

Soldiers' thoughts as they moved through the scenic late-summer landscapes on their way to the front were probably as diverse as the lives they left behind. To some, it might initially have seemed like a vacation, especially since for many it was the first time they had left their hometown or village.[15] Farmers might have been reminded of the impending harvest at home. Others, particularly those who had participated in youth organizations such as the *Wandervögel* in Germany or Britain's Boy Scouts, might have viewed the coming war as an outdoor adventure that offered escape from the constraints of their parochial lives and bourgeois societies.[16] Yet no one, not even the military leadership, could predict what actually lay ahead at the western front, because battlefields were determined as much by weather conditions and topographical features as by tactical considerations and strategic imperatives.

In its first month World War I was still a war of movement. Following their invasion of Belgium on the fourth of August, four German armies marched toward Paris at a pace of twenty to thirty kilometers a day. Whereas initially some soldiers might have appreciated sleeping under the stars and marching through the countryside, they soon realized that going to war was not a romantic outdoor exploit but a physically taxing ordeal, requiring long marches with heavy knapsacks in often ill-fitting boots.[17] Marc Bloch, a young French soldier who would later become a founder of the Annales School, described his first impressions of war in August 1914 as "beautiful days, very calm but a bit monotonous.... The sun, the rustic pleasures—fishing, swimming in the river, and dozing on the grass"; just a few days later, however, he "realized that this hope was misplaced. This immensely bitter disappointment, the stifling heat, the difficulties of marching along a road encumbered by artillery and convoys, and finally, the dysentery with which I was stricken the night before made the 25th of August ... one of the most painful days I have known."[18]

As soldiers made the transition from home to the front lines, they learned that going to war was not about hiking through a landscape but about surviving an experience that bore little resemblance to a serene stroll in the country. While most of these soldiers were crossing territory unknown to them, they were nevertheless passing through a recognizable landscape, one that had not yet been exposed to the destructive force of modern military technology. In the coming four years, soldiers would experience at firsthand how these landscapes of peace became environments of war; these men were not merely observers of but active participants in this process of transformation, especially once the armies reached a stalemate following the battle of the Marne in early September 1914.

Landscapes of Entrenchment

While not an invention of World War I, trench warfare became synonymous with the western front.[19] The first trenches were dug in September 1914 on the northern banks of the Aisne River and were considered a transitory phenomenon in a war of movement. Hastily constructed, they served mainly as rifle pits and offered little more cover than a ditch; but as summer turned to autumn, it became clear that neither would the war end quickly nor would the trenches be temporary. On the contrary, over the next four years they would develop into a network of trenches more than five hundred kilometers long and reaching from the Swiss border to the Baltic Sea.[20] Their specific location depended on the enemy's position as well as the topography. The flat fields stretching from Flanders all the way to the Forest of the Argonne, for example, offered conditions favorable for building trenches because below the top layer of humus and clay, the ground consisted mainly of easy-to-dig chalk (figure 4.1).

Each country developed a distinctive style of trench building, which, especially in the case of Britain and Germany, became increasingly complex as the conflict wore on.[21] Whereas the trenches initially consisted of a single row on each side, in time they expanded into a multisector system composed of firing, support, and reserve lines, which were linked by perpendicular com-

Fig. 4.1 Battlefield in the Argonne Forest, 1916. Unknown photographer. © *Bildarchiv Preußischer Kulturbesitz.*

Environments of Death

Fig. 4.2 Trenches and bunkers, 1916. Postcard, unknown photographer. Courtesy of the Mauch and Roller families, Weil im Schönbuch, Germany.

munication trenches and outfitted with strong points, gun posts, dugouts, and latrines.[22] Some German firing trenches even contained several levels of dugouts for soldiers and officers, complete with staircases molded into the wall of the trenches that in some cases descended as far as ten meters below the surface (figure 4.2). Trench walls were reinforced, and wooden planks, or duckboards, lined the floors. Nevertheless, trenches always remained provisional, not least because they could be destroyed at any moment by a direct hit from enemy artillery.

A tour in the firing trenches was supposed to last between three and five days, but soldiers frequently had to remain in them for weeks on end. Having to live, eat, sleep, and fight in a trench that was only a few feet wide, soldiers' day-to-day life was reduced to manning weapons, building and maintaining trenches and communication lines, and on occasion going over the top. Frontline duty alternated between heart-pounding excitement and crushing boredom; and through it all soldiers lived literally in the dirt.

These conditions did not lend themselves to the chivalrous existence of the heroic warrior but rather, as Louis Mairet put it, to "a life so frightfully bestial ... even pigs are better off!"[23] Indeed soldiers often referred to themselves as *Frontschwein* or *poilu*. Wolfgang Mommsen aptly summed up the situation when he wrote: "The conventional life of soldiers was reduced to the daily struggle against the cold, mud, and wetness, against illnesses of all kinds and having to endure the massive shrapnel and artillery attacks without being able to do anything about it except to dig deeper and to make one's own positions as bulletproof as possible."[24] Soldiers spent much of their time repairing trenches, mostly under the cover of darkness so as to avoid being targets for enemy snipers.

In addition to human labor, the building and maintenance of these trenches required a considerable amount of natural resources, especially wood. Consequently, large segments of French and Belgian forests were cut down in order to supply the materials for trench construction.[25] But logging marked only the beginning of a much greater transformation of the landscape wrought by techniques of modern warfare.

Entrenchment created a new spatiality above and below ground, which altered both the conduct and the experience of war. For one, the battlefields of World War I no longer offered a panoramic view, as those of previous wars had. On traditional battlegrounds, generals had surveyed a vista that helped them to plan assaults. In World War I the army commands rarely saw the front lines, which not only made military strategies increasingly abstract, but also created a growing rift between those strategies and subsequent battle tactics.[26] To many commanders, battlefields continued to be landscapes transposed onto maps, which may be one reason why they were able to hold on to conventional notions of offensive warfare as long as they did.[27] It was easy to send men "over the top" because from the generals' perspective, soldiers were still traversing a landscape rather than holding a stationary position in an elaborate network of trenches obstructed by debris, barbed wire, corpses, and fallen trees.

Tellingly, soldiers on the attack were sent into a space called "no-man's-land," the precinct between the opposing lines of trenches that could be anywhere from four to a thousand meters wide. Serving as the actual battleground on which hand-to-hand combat took place, no-man's-land grew into a cratered, debris-strewn arena of destruction that was supposed to be conquered and filled in but that could in fact barely be inhabited.[28] It signified the reversed spatiality of the combat zone, where the living had to stay hidden below ground while corpses gradually claimed the space above ground.

This reversal of space was largely due to new technologies.[29] World War I witnessed the emergence of a whole arsenal of new weapons, including rapid-fire machine guns and several types of field and heavy artillery, as well as airplanes, flamethrowers, tanks, and poison gases.[30] In light of these innovations, traditional notions of battlefields, as Cornelia Vismann has argued, had to be reconceptualized.[31] Due to the increased range of some of these weapons—a British Howitzer, for example, could hit a target ten kilometers away—"front lines" became "battle zones" that were not only several kilometers wide but also reached underground and into the sky.[32] Moreover, this new technology generated a capacity for destruction that no longer focused just on the killing of individual soldiers; now warfare also included the obliteration of entire landscapes.[33] Heavy artillery and the shrapnel it produced gouged and cratered the earth, destroying soldiers, animals, trees, drainage systems, and communication lines without distinction. This unprecedented destruction was, to some extent, an unintended consequence of the deployment of high-powered explosives in a stalemate war (*Stellungskrieg*). It demonstrated how machine warfare created an arena of indiscriminate destruction. At the same time, new weapons, most notably aircraft, also offered a new visual perspective on the front, namely a view from above. Interestingly, just as the overview of the ground was lost, reconnaissance airplanes began to generate a panoramic view of the battlefield from above.[34] Yet the use of airplanes also forced soldiers farther below ground so that their positions would be less visible.[35]

In trench warfare, humans were subordinated to the environment. As one captain in the Austrian cavalry wrote to his son, "Modern combat is played out almost entirely invisibly, the new way of fighting demands of the soldier that he withdraw from the sight of his opponent. He cannot fight upright on the earth but must crawl into and under it."[36] One characteristic of the modern warrior was his ability to blend in with the landscape and become indistinguishable from the environment. Military uniforms were an obvious indicator of this changing relationship between soldiers and their environment. Whereas in previous wars brightly colored uniforms had served to distinguish soldiers from the enemy and the environment, soldiers now tried to blend in by wearing more subdued earth tones like brown and green.[37] Initially, the French troops continued to wear traditional uniforms, but it quickly became apparent that red trousers and blue coats were a liability because they turned soldiers into easy targets for German snipers. Even military helmets had to be redesigned to better suit the circumstances of trench warfare. For example, in 1916, the German *Pickelhauben* were replaced by camouflaged steel helmets.

Perceptions in an Environment of War

Trench warfare forced soldiers to develop a new relationship with space, including intensified sense perceptions.[38] To some soldiers, going to war might initially have looked like an adventure, but they quickly realized that life at the front was nothing like tourism. For one thing, there was little to see. Trench warfare no longer privileged sight, particularly when it came to locating the enemy. Not only was the landscape increasingly unrecognizable due to military destruction; most soldiers spent large amounts of time close to or even below ground, where their field of vision was limited to the boundaries of the trenches, creating a peculiar perspective. As a result, battlefields looked empty even though they were actually saturated with bodies, both living and dead.[39] Even inside the trenches, soldiers often could not see very far because of the trenches' zigzag construction. The view across no-man's-land was obstructed by barbed wire and upturned earth, and during a barrage this field of vision was even further reduced when smoke or poison gas filled the air.

In the absence of sight, sounds took on greater significance. Trying to make sense of these hitherto unknown sounds, some at the front, including the expressionist painter-turned-soldier Franz Marc, resorted to comparisons with natural phenomena. Marc described the soundscape of battle as "stormlike cannon thunder."[40] "The thunder of the cannons," another soldier reported, "is often so lively that one cannot hear a single cannon shot but only an uninterrupted rumble that goes on for hours."[41] This noise, which became known as *Trommelfeuer* (literally "drumfire"), was often so intense that it could be heard even in the rear areas several kilometers behind the front lines.

One of the first things soldiers had to learn upon arriving at the front was how to interpret sounds as indicators of danger.[42] The noise of barrages was initially disorienting for most soldiers. Of his early days at the front, Ernst Jünger wrote: "still unfamiliar with the sounds of war, I was unable to disentangle the whistle and hissing, the banging of our own guns and the cracking of enemy grenades that hit at ever-smaller intervals."[43] He spoke of the need to identify sounds in order to "create an image out of all this."[44] Trench life was, in many ways, a synesthetic experience.

According to Mary Habeck, soldiers learned first to differentiate among the many sounds heard during an attack and, second, to tolerate the constant barrages.[45] After a while most soldiers could recognize the warning each type of sound represented. As a Canadian infantryman wrote: "One could identify it from the pitch of the noise where the grenade would hit. If it sounded like a moving train, the target was several kilometers away, a shrieking sound or deep rumble meant immediate danger."[46] Even though soldiers learned how to differentiate among sounds, sometimes "the thunder of battle was so terrible that no one was of sound mind anymore," especially since, in addition to the noise of barrages, soldiers had to endure the cries of wounded men and animals.[47]

Whereas soldiers could learn to distinguish sounds, when it came to smells, they simply had to cope with the stench. Given the unhygienic conditions at the front lines—the lack of washing facilities and latrines so noxious that most soldiers preferred not to use them—the odors must have been overpowering. Moreover, the battlefield was marked by the smell of death. Decaying human and animal flesh in particular contributed to the effervescing stench, which only intensified as the war dragged on because an increasing number of the dead could no longer be buried properly, if at all. Reflecting on this penetrating stink, Marc Bloch wrote: "The stench turns one's stomach since the ground was strewn with all sorts of debris, weapons, equipment, and human fragments."[48] Similarly, Ernst Jünger noted:

> A thick smell of dead bodies lay over the ruins, as on all dangerous zones of this area, because the fire was so strong that no one could take care of the casualties. One truly ran for one's life, and as I took in this smell while running, I was hardly surprised—it belonged to the place. Not only was this heavy and sweet odor repulsive, but mixed

with the stinging fog of explosives, it caused a kind of excitement that can be created only by the immediate presence of death.[49]

Whereas for Jünger this odor of death heightened the excitement of war, others, such as Franz Marc, felt that "the stench of corpses across many kilometers is the worst. I can bear it less than seeing dead people and horses."[50] The smell of death attached itself to everything. At the Battle of Verdun, one Frenchman noted, "We all had on us the stench of dead bodies. The bread we ate, the stagnant water we drank, everything we touched had a rotten smell."[51] Survival depended on filtering out some sensory impressions while remaining hypersensitive to others, developing immunity to the most shocking sights, sounds, and smells of war while staying alert for any sign of danger. Soldiers unable to regulate their senses were more likely to develop shell shock and ran a higher risk of becoming a casualty.[52] The environment of the western front at least temporarily unhinged conventional notions of order with respect to sensory experiences and conceptions of space.[53] World War I turned into a total war because it ultimately demanded the mobilization of all sectors of society. Moreover, it was experienced as all consuming by those involved, owing to the total destruction caused by a style of fighting that had replaced man-to-man combat with warfare against everything.

Trench warfare not only shaped the daily perceptions of soldiers, it also altered the composition of the land. The scenic landscapes of the western front had disappeared by November 1914, a transformation that the newly emerging front-line photography amply documented.[54] Already in late September of that year Marc Bloch and his fellow soldiers had dug a trench in the woods of La Gruerie, but when they returned the next day, they found only a clearing in place of the underbrush because the Germans had continued to shell the position long after the French abandoned it.[55] A letter of November 1914 from a German soldier at the front reported "all the trees shot up, the earth gouged by heavy artillery, and then animal corpses, destroyed houses and churches, nothing, nothing was even close to being useable anymore. And every unit that has to advance must move for many kilometers through this chaos, this stench of corpses and the huge mass grave."[56]

In the storm of steel, vegetation disappeared and was replaced by a different mixture of soil, body parts, and military debris. As a result, soldiers easily lost their sense of direction in the maze of trenches and torn-up earth. Trench warfare required new markers for orientation, some of them quite harrowing. For example, one British soldier remarked: "we learned the landmarks to guide us: 'Left by the coil of wire, right by the French legs.'"[57] Describing the environment of the Battle of Loos, a British soldier in the 21st Division wrote: "The ground was so uneven that headway was difficult to make, not uneven by nature either, but by the huddled heaps of men's bodies."[58] Another soldier observed, "this ground was full of corpses, mostly French, and when we dug, we resurrected bits of them."[59]

The western front gave rise to an environment where men, animals, machines, trees, and the earth were reconfigured into an unrecognizable landscape of total destruction in which death became visible while the living had to stay hidden in order to survive. According to Bernd Hüppauf, "World War I provided the first image of the total destruction of landscape and an example of the unlimited dominance of technology over nature and space."[60] While trench warfare attested to the destructive human power over nature, it also underscored how humans were both continuously subordinated to natural forces and subjected to the unintended consequences of this environmental destruction, because in the trenches of the western front, as in all wars, the environment also became an additional threat.

Belligerent Environments

Enemy fire was not the only wartime threat. Front-line soldiers also had to deal with changing weather conditions. Each season brought its own set of hazards that affected the day-to-day conduct of war—snow and freezing temperatures in the winter; heat, mosquitoes, and stench in the summer. On occasion, some soldiers even seemed to prefer combat to being exposed to the weather. For example, in early 1915, one French soldier wrote: "We don't think of death, but it's the cold, the terrible cold! It seems to me at the moment that my blood is full of blocks of ice. Oh, I wish they'd attack, because that would warm us up a little."[61] Soldiers in the opposing trenches expressed the same sentiment. Christian Krull, for example, noted, "We could not sleep because of all the vermin, wet, and cold. We went to the artillery gun and replaced the deadly tired gunners just so that we could get warm."[62] When it came to enduring these environmental conditions, all nations suffered equally.

One of the biggest problems, however, was the incessant rainfall that transformed the front into a sea of mud. It was one of the complaints soldiers voiced most often in their letters and in the front-line press. Describing the first day of June 1916, the French trench paper *L'Argonnaute* reported, "It has been raining all day, that cold, fine, relentless winter rain, against which there is no protection. The front-line trench is a mud-colored stream, but an unmoving stream where the current clings to the banks of its course. Water mud. You go down into it, you slip in gently, drawn in by who knows what irresistible force.... Everything disappears into this ponderous liquid: men would disappear into it too if it were deeper."[63] Sometimes, the rain was so severe that "though we were well aware of the shells, we really thought only of the rain. Inexhaustible clouds dumped an almost incessant downpour on the underbrush. The clay soil held the water on its surface. Our trenches were a brook, the woodland roads were lakes of mud, and the ditches alongside were tumbling torrents of a yellowish flood."[64] Indeed, during heavy downpours soldiers spent much of their time pumping water and mud out of the trenches. At times, torrential rains actually affected battles. For example, January 2, 1915, was apparently such a "dreadful day of rain, everything is so hazy that one cannot think about shooting."[65] Fog and continuous rainfall delayed the start of the Battle of Verdun for nine days.[66]

In addition to rain, mud became a chronic nuisance. In the words of one British soldier: "The memory of that [i.e., the western front] is mainly—mud."[67] Marc Bloch even referred to World War I as "the age of mud."[68] The chalky soil of Flanders and northern France had been easy to dig out, but "when the rain came in autumn, the trenches disappeared and the area became a lake of mud."[69] After days of rain, soldiers found that, "When we reach the communication trench, it is no longer a trench at all but a stream of fluid mud, where we sink over our leggings. We have to use our hands to pull out our legs when they get stuck."[70] And mud was not just unpleasant; it contributed to the hazards of front life. For example, one British soldier complained, "the mud on my greatcoat made it monstrously heavy, so that it flapped like lead against my legs, making the going utterly wearisome. I would willingly have died just then."[71] According to a March 1917 article in the French trench newspaper *Le Bochophage*, "mud throws its poisonous slobber out at him, closes around him, buries him.... For men die of mud, as they die of bullets, but more horribly. Mud is where men sink and—what is worse—where their souls sink."[72] Some even compared the mud to an "enormous octopus" that swallowed men up, insisting, "hell is not fire, that would not be the ultimate in suffering. Hell is mud!"[73]

Another harrowing aspect of daily life in the trenches was the constant presence of vermin, most notably lice and rats. Weather conditions and vermin had always been a part of war to some degree. Whereas rain and mud underscored the continuous impact of weather, vermin pointed to the environmental consequences of stalemate warfare, which offered favorable conditions for the rats and lice that fed off human bodies, debris, and food supplies. The omnipresent vermin intensified soldiers' everyday misery, heightening the sense that civilization was coming to an end. One French trench paper described the situation as follows: "We would be thrilled not to find our bread contaminated by rats and our shirts invaded by vermin. If someone says, 'the *boches* are 20 meters away,' we feel a chill; but if we are told: 'the dug-outs are full of lice'—that we find really disgusting!"[74] Soldiers spent much of their time "chatting," mainly because to chat, at least in British parlance, meant to pick lice out of one's uniform (a "chat" was a louse); it proved to be a rather futile task, however, since it was ridiculous to kill an individual louse if one had hundreds of them.[75]

Whereas lice clung to soldiers' bodies, rats infested their surroundings. For rats, the trenches provided ideal habitats in which food was abundant and natural enemies were scarce (apart from soldiers and their rat-fighting pets). Feeding on food rations and the corpses of dead humans and animals, rats thrived and multiplied at an astonishing rate. Oddly enough, enemy fire was one of the few things that kept rat populations in check. Soldiers frequently reported how armies of rats fled areas where a grenade had hit.[76] Gas, too, tended to create panic among rat populations, which often served as early warning of an impending attack. A lieutenant colonel in the London Regiment at the Somme described a ghastly scene where, in the aftermath of a gas attack, "The trenches swarmed with rats, big rats, small rats, grey rats, tall rats in every stage of gas poisoning! Some were scurrying along scarcely affected while others were slowly dragging themselves about trying to find a corner in which to die. A most horrid sight—but a very good riddance."[77]

Apart from adding to the repulsive nature of life at the front, vermin also posed direct threats to soldiers' health. For example, just before the war, scientists had discovered that lice spread the bacterium that causes typhus, a highly lethal contagious disease that debilitated thousands of front-line soldiers and killed a fourth of those afflicted.[78] Rats also carried diseases that could be spread to humans. In general, the damp conditions coupled with poor hygiene in confined spaces increased soldiers' susceptibility to viral and bacterial infections of all kinds. The French army alone recorded close to five million cases of contagious disease between 1914 and 1918.[79] Even though World War I was the first military conflict in which more casualties were caused by weapons than disease, epidemics and other ailments continued to be a significant threat.[80] An ailment specific to World War I was trench foot, which occurred due to "prolonged standing in cold water or mud and by the continued wearing of wet socks, boots, and putters."[81] Even more perilous was the emergence of a new form of trench fever, which was first diagnosed in 1915.[82] Here, too, the louse proved to be the main vector, although its role in transmission was not established until after the war. To make matters worse, in 1917 influenza began to spread at the front, affecting 708,306 men on the German side alone between August 1917 and July 1918.[83] By that time, however, disease was far from the greatest threat to soldiers at the front.

On April 22, 1915, the Germans had launched the first successful chemical-weapons attack of the war, deploying more than 160 tons of chlorine gas at the Second Battle of Ypres.[84] This event marked a turning point in the history of military technology and inaugurated a type of warfare in which the environment itself could become a lethal weapon.[85] Conventional methods of combat were no longer the only or even the most efficient ways to wage war. Instead of penetrating the enemy's body with bombs, bullets, or blades, poison gas destroyed an adversary from the inside out.

Over the course of the war, all sides resorted to the use of poison gas. The British began using chlorine gas by September 1915, and by the end of the war, they had conducted 768 gas attacks discharging 57,000 tons of gas.[86] The number of gas shells rose from 1 percent of all fired shells in 1916 to 30 percent by 1918 in some sectors of the front.[87] The primary purpose of using gas was to terrorize the enemy. At first, gas was mainly utilized to initiate an attack, but with the arrival of chlorine shells, gas became a deadly weapon of attack that unleashed its yellow or greenish clouds over enemy trenches, blinding and suffocating soldiers en masse. As the French trench paper *Le Filou* stated: "We had seen everything: mines, shells, tear-gas, woodland demolished, the black tearing mines falling in fours, the most terrible wounds and the most murderous avalanches of metal—but nothing can compare with this fog which for hours that felt like centuries hid from our eyes the sunlight, the daylight, the clear whiteness of the snow."[88] In a similar vein, a British doctor remarked:

> I shall never forget the sights I saw up by Ypres after the first gas attack. Men lying all along the side of the road between Poperinge and Ypres, exhausted, gasping, frothing yellow mucus from their mouths, their faces blue and distressed. It was dreadful and so little could be done for them. One came away from seeing or treating them longing to be able to go straight away at the Germans and to throttle them, to pay them out in some sort of way for their devilishness. Better for a sudden death than this awful agony.[89]

Gas was harmful not only to humans; it adversely affected every living creature. As *Le Filou* reported on March 20, 1917: "Little birds fell into the trenches, cats and dogs, our companions in misfortune, lay down at our feet and did not awaken."[90] Ernst Jünger, too, recalled how, after a gas attack, "A large number of plants were wilted, snails and moles were lying dead, and the messenger horses that were stabled in Monchy had water running out of their eyes and mouths."[91]

Gas represented a new weapons technology that did not kill directly but altered the environment in such a way as to make it uninhabitable. In order to comprehend the full extent of the devastation caused by World War I, we must recognize that machine-age warfare was not simply the result of other technological means to kill people but also of incorporating the environment into this destruction. In other words, when thinking about the place of the environment in warfare, we should think of it not merely as a stage for combat but as a constitutive element of warfare that directly influenced the course of battles. For one, weapons functioned only if the environment cooperated. In World War I this was often not the case, as was shown by, for example, the inability of tanks—another innovation of this war—to move through the cratered fields of mud. Hence, neither environment nor weaponry can be viewed as completely separate entities because both were interdependent in the age of total warfare (figure 4.3).

World War I required a new synergy not just between man and machines but also between technology and the environment, including animals. During World War I the interrelationship of weaponry and the environment reached an unprecedented level in the sense that innovations in military technology, particularly heavy artillery and poison gas, led to a type of destruction that no longer focused exclusively on the killing of soldiers but instead aimed at the

Fig. 4.3 Devastated landscape near Ypres with disabled British Mark IV tank, 1917. Unknown photographer. © Bildarchiv Preußischer Kulturbesitz.

indiscriminate and total annihilation of everything. It was modern technology that made possible the mass slaughters that characterize twentieth-century warfare.

Conclusion

By November 1918, four million of the approximately thirty-two million men in uniform had been killed, and millions more had been permanently disabled or disfigured.[92] At the Battle of the Somme alone, France lost every third soldier of the 1.5 million they had sent in. Of the 2.5 million Germans, 700,000 became casualties of war. The British army lost more men in these three months than during the entirety of World War II.[93] In addition to the human death toll, millions of animals were killed in this stalemate slaughter.

Villages and towns lay in ruins, fields had been turned into moonscapes, and forests had been reduced to acres of stumps (figure 4.4). The French Forestry Service estimated that 350,000 hectares of forest had been destroyed during the war, an amount that would have supplied the tree harvest for the next sixty years.[94] As the writer Arnold Zweig describes it: "Beautiful villages first became ruins, then mounds of rubble, and finally stretches of bricks; woods first [became] gaps and tangles, then fields of dead pale stumps, eventually desert."[95]

Environments of Death

Fig. 4.4 Tree ravaged by artillery fire, 1915. Postcard, unknown photographer. Courtesy of the Mauch and Roller families, Weil im Schönbuch, Germany.

The conflict had also unsettled existing notions of landscape and property relations, imposing its own set of war-specific environmental conditions, which grew out of the intended and unintended consequences of military destruction. The western front gave rise to a distinct environment of trenches and no-man's-land filled with the intense sounds, sights, and smells of a war of attrition bent on the massive destruction of human lives as well as nature. Indeed, symbolically the death of soldiers was closely tied to the death of nature. One manifestation of this connection between soldiers and nature was the commemoration of the dead. Describing the German *Heldenhaine,* George Mosse wrote: "Nature itself was supposed to serve as a living monument; the German forest provided a fitting environment for the cult of the fallen soldier."[96]

War created a new *Schicksalsgemeinschaft* (community of fate) between men and nature, who threatened each other in the daily practice of warfare, but who also amalgamated into a new symbolic unity born out of their mutual annihilation. According to Ernst Toller, "A forest is a Volk. A shot-up forest is an assassinated Volk."[97] Even more tellingly, another German soldier wrote: "The fate of this forest is linked and interwoven with my own at the deepest level. Not only were the woods my comrade but also my protection, a shield that wards off the lead and iron hurled at me . . . allowing itself be pierced to the heart; it yields to death without resistance so that I may live. Often when thick clouds have obscured the sun . . . [and] the day is full of melancholy and sadness, then bright drops trickle from the trees' flayed crowns like tears of never-ending pain."[98]

Notes

The epigraph is taken from "In Stahlgewittern," in *Sämtliche Werke,* 18 vols. (Stuttgart: Klett-Cotta, 1978–83), 1:44: "Das Gesicht der Landschaft war finster und fabelhaft, der Kampf hatte das Liebliche der Gegend hinweggewischt und seine ehrenden Züge hineingegraben, vor denen der einsame Betrachter erschrak." Unless otherwise noted, translations from the original German are my own.

1. Kurt Lewin made this distinction between landscapes of peace and war in his 1916 essay "Kriegslandschaft," *Zeitschrift für Angewandte Psychologie* 12 (1917): 440–47.

2. Denis Cosgrove, "Landscape and Landschaft," *Bulletin of the German Historical Institute, Washington, D.C.* 35 (Fall 2004): 57–71.

3. On the linkages between landscape and culture, see, in addition to the abovementioned article by Denis Cosgrove, his "Landscape as Cultural Product," in *Theory in Landscape Architecture: A Reader,* ed. Simon Swaffield (Philadelphia: University of Pennsylvania Press, 2002), 165–66; see also John Brinckerhoff Jackson, *Discovering the Vernacular Landscape* (New Haven, Conn.: Yale University Press, 1984).

4. Cosgrove, "Landscape and Landschaft," 61.

5. Ibid., 68.

6. I do not mean to imply that landscapes are necessarily picturesque. Surely every war also has a landscape, even if it is marked by destruction. Nor am I suggesting that landscapes are destroyed only during wars; quite the opposite, their destruction is often more severe in times of peace due to agricultural exploitation and land development.

7. It would certainly be equally interesting to investigate other World War I front lines in the east or the Alps as well as the sea (and I hope someone will). My focus on the western front is based on the manageability of this topic rather than a claim of exceptionalism.

8. "Seit zehn Tagen haben wir zum ersten Mal dauernden Sonnenschein, da läßt sich doch alles noch ertragen, aber grausig waren die ersten Wochen mit ihrem unaufhörlichen Regen, der alles ringsum in Sumpf verwandelt. Man konnte keine zehn Schritt weit gehen, ohne daß man bis an die Knie versank." Letter from Christian Krull dated July 29, 1916, quoted here from *Frontalltag im Ersten Weltkrieg: Wahn und Wirklichkeit: Quellen und Dokumente,* ed. Bernd Ulrich and Benjamin Ziemann (Frankfurt: Fischer, 1994), 90.

9. One exception is Shepard Krech III, J. R. McNeill, and Carolyn Merchant, eds. *The Encyclopedia of World Environmental History,* 3 vols. (New York: Routledge, 2004), which includes an entry on warfare written by Richard Tucker (3:1284–91). For an introduction to the existing literature, see Richard P. Tucker and Edmund Russell, eds., *Natural Enemy, Natural Ally: Toward an Environmental History of War* (Corvallis: Oregon State University Press, 2004). In their introduction, Tucker and Russell remark that one of the biggest challenges was to actually find enough contributors for such a volume.

10. Tony Ashworth, *Trench Warfare, 1914–1918: The Live and Let Live System* (London: Macmillan, 1980); Stéphane Audoin-Rouzeau, *Á travers leurs journaux: 14–18 Les combattants des tranchées* (Paris: Armand Colin, 1986); Gerhard Hirschfeld, ed., *Kriegserfahrungen: Studien zur Sozial- und Mentalitätsgeschichte des Ersten Weltkriegs* (Essen: Klartext, 1997); Eric J. Leed, *No Man's Land: Combat and Identity in World War I* (Cambridge: Cambridge University Press, 1979); Frédéric Rousseau, *La Guerre censurée: Une histoire des combattants européens de 14–18* (Paris: Seuil, 1999); Ulrich and Ziemann, eds., *Frontalltag im Ersten Weltkrieg;* Richard Van Emden, *The Trench: Experiencing Life on the Front Line, 1916* (New York: Bantam, 2002); and Jay Winter, *The Experience of World War I* (London: Macmillan, 1988).

11. Philippe Dagen, *Le silence des peintres: Les artistes face à la Grande Guerre* (Paris: Fayard, 1996); Modris Eksteins, *Rites of Spring: The Great War and the Birth of the Modern Age* (Boston: Houghton Mifflin, 1989); Paul Fussell, *The Great War and Modern Memory* (New York: Oxford University Press, 1975); Bernd Hüppauf, "Räume der Destruktion," *Krieg und Literatur* 3 (1991): 105–23; Annegret Jürgens-Kirchhoff, "Verbrannte Erde: Kriegslandschaften in der Kunst zum Ersten und Zweiten Weltkrieg," in *Erster Weltkrieg, Zweiter Weltkrieg: Ein Vergleich: Krieg, Kriegserlebnis, Kriegserfahrung in Deutschland,* ed. Bruno Thoß and Hans-Erich Volkmann (Paderborn: F. Schöningh, 2002), 783–819; Sue Malvern, *Modern Art, Britain and the Great War: Witnessing, Testimony, and Remembrance* (New Haven, CT: Yale University Press, 2004); Linda McGreevy, *Bitter Witness: Otto Dix and the Great War* (New York: Peter Lang, 2001); and George L. Mosse, *Fallen Soldiers: Reshaping the Memory of the World Wars* (New York: Oxford University Press, 1990).

12. The initial mobilization of the Belgians amounted to 177,000 soldiers. Britain mobilized 733,500, France 3.781 million, and Germany sent 4.5 million soldiers to the front in 1914. John Ellis and Mike Cox, eds., *The World War I Databook: The Essential Facts and Figures for All the Combatants* (London: Aurum, 2001), 245.

13. In World War I, the front-line mobility of most European armies still depended primarily on horsepower for the transport of weapons, military supplies, food rations, and ultimately wounded and dead soldiers. Moreover, the strategic thinking of most generals, at least initially, continued to be rooted in traditional notions of cavalry attacks—a strategy that quickly proved devastating, particularly for the British military. The German cavalry alone brought 715,000 horses to the front. The British army started with 40,000 animals,

but by the end of the war, 404,000 animals had been sent to the front. J. M. Bourne, *Britain and the Great War, 1914–1918* (London: Edward Arnold, 1989), 177. In addition to horses, messenger dogs and pigeons were increasingly used to keep up communication between different command posts, especially when bombardments had destroyed telephone lines. The role of animals in war has hardly been studied. To the extent that it has been investigated, most studies tend to be anecdotal or numeric rather than analytical. See Ernest Baynes, *Animal Heroes of the Great War* (New York: Macmillan, 1925); Michael G. Lemish, *War Dogs: A History of Loyalty and Heroism* (Washington, D.C.: Brassey's, 1999); Robert E. Lubow, *The War Animals* (Garden City, N.J.: Doubleday, 1977); Martin Monestier, *Les animaux-soldats: Histoire militaire des animaux des origins à nos jours* (Paris: Le Cherche Midi, 1996); John Singleton, "Britain's Military Use of Horses, 1914–1918," *Past & Present* 139 (May 1993): 178–203.

14. It was a conviction that would prove fatal, both politically and strategically, not the least due to the changing weather conditions, most notably the arrival of winter, which found many troops unprepared. On the initial attitudes toward war, see, for example, Jean Jacques Becker, *1914: Comment les Français sont entrés dans la guerre* (Paris: Presses de la Foundation nationales des sciences politiques, 1977); Ian Beckett, *The Great War, 1914–1918* (London: Longman, 2001), 42–55; Thomas Rohkrämer, "August 1914—Kriegsmentalität und ihre Voraussetzungen," in *Der Erste Weltkrieg: Wirkung, Wahrnehmung, Analyse*, ed. Wolfgang Michalka (Munich: Piper, 1994), 759–77; and Jeffrey Verhey, *The Spirit of 1914: Militarism, Myth, and Mobilization in Germany* (Cambridge: Cambridge University Press, 2000).

15. See, for example, Sabiene Autsch, ed., *Der Krieg als Reise: Der Erste Weltkrieg—Innenansichten* (Siegen: C. Böschen, 1999).

16. Eksteins, *Rites of Spring*, 133; and Hüppauf, "Räume der Destruktion," 115–16.

17. Bernd Ulrich, "Die Desillusionierung der Kriegsfreiwilligen von 1914," in *Der Krieg des kleinen Mannes: Eine Militärgeschichte von unten*, ed. Wolfram Wette (Munich: Piper, 1992), 110–26.

18. Marc Bloch, *Memoirs of War, 1914–1915*, trans. Carole Fink (Ithaca, N.Y.: Cornell University Press, 1980), 80–81.

19. On the emergence of trench warfare, which actually originated in the American Civil War and the Russo-Japanese conflict, see Ashworth, *Trench Warfare*, 2–23; and Habbo Knoch, "Die Front," in *Orte der Moderne: Erfahrungswelten des 19. und 20. Jahrhunderts*, ed. Alexa Geisthövel and Habbo Knoch (Frankfurt: Campus, 2005), 270–80. For a more contemporary perspective, see André Laffarque, *The Attack in Trench Warfare* (New York: Van Nostrand, 1917); Francis Marre, *Dans les tranchées du front* (Paris: Bloud and Gay, 1915); *Notes for Infantry Officers on Trench Warfare Compiled by the British General Staff* (Washington, DC: U.S. Army, 1917), 19–22; and Joseph Shuter Smith, *Trench Warfare: A Manual for Officers and Men* (New York: E. P. Dutton, 1917).

20. Indeed, warfare became so identified with the trench experience that a growing number of model trenches were put on exhibit throughout Germany and Britain in order to give those who were still at home a sense of life in the trenches, albeit a sanitized one. Britta Lange, *Einen Krieg ausstellen: 'Die deutsche Kriegsausstellung' 1916 in Berlin* (Berlin: Verbrecher, 2003); and Eva Zwach, *Deutsche und englische Militärmuseen im 20. Jahrhundert: Eine kulturgeschichtliche Analyse des gesellschaftlichen Umgangs mit Krieg* (Münster: Lit, 1999).

21. The French army kept their trenches simple, in part because they considered entrenchment a sign of the Germans' cowardice to fight in the open. Leonard Smith, Stéphane Audoin-Rouzeau, and Annette Becker, *France and the Great War, 1914–1918* (New York: Cambridge University Press, 2003), 89.

22. Latrine construction posed particular challenges because these facilities had to serve a large number of men. One manual suggested that latrines should be located about forty feet behind the fire trenches off the communication trenches. The pit should be twelve feet deep and long, and covered with boards so that as many soldiers as possible could use them at any one time. To prevent rat and fly infestations, disinfectants were to be applied generously, and once the latrines were filled to six feet, they were to be replaced. Because they were often mistaken for gun posts and targeted by enemy artillery, they also had to be concealed. See Smith, *Trench Warfare: A Manual*, 27–28.

23. Louis Mairet, *Carnet d'un combattant* (Paris, 1919), 96 (diary entry for September 27, 1915).

24. "Das herkömmliche Soldatenleben wurde weitgehend reduziert auf den täglichen Kampf gegen Kälte, Schlamm und Nässe, gegen Krankheiten aller Art und auf das massive Erleiden von Schrappnel- und Artilleriebeschuß, ohne etwas dagegen tun zu können, als sich immer tiefer einzugraben und die eigenen Stellungen möglichst beschußsicher auszubauen." Wolfgang Mommsen, *Die Urkatastrophe Deutschlands: Der Erste Weltkrieg, 1914–1918* (Stuttgart: Klett Cotta, 2002), 123.

25. Andrée Corvol, ed., *Guerre et Forêt* (Paris: Éditions L'Harmattan, 1994); and Richard P. Tucker, "The World Wars and the Globalization of Timber Cutting," in Tucker and Russell, eds., *Natural Enemy, Natural Ally*, 110–41.

26. Michael Howard, "Men against Fire: Expectations of War in 1914," in *Military Strategy and the Origins of the First World War*, ed. Steven E. Miller, Sean M. Lynn-Jones, and Stephen Van Evera (Princeton, N.J.: Princeton University Press, 1991), 3–19.

27. One of the things I want to investigate further is the extent to which environmental factors played into the planning of military campaigns on the western front.

28. Oddly, despite its destruction, no-man's-land also offered favorable conditions for some creatures and plants. The story of the poppies, which became an enduring symbol of World War I especially in Britain, is well known. One of the few creatures that seemed to thrive in no-man's-land was partridge because they had lost their natural enemies.

29. On the technological changes in World War I, see, for example, Mary Habeck, "Die Technik im Ersten Weltkrieg—von unten gesehen," in *The Great War and the Twentieth Century*, ed. Jay Winter, Geoffrey Parker, and Mary Habeck (New Haven, Conn.: Yale University Press, 2000): 101–32; Hubert C. Johnson, *Breakthrough! Tactics, Technology, and the Search for Victory on the Western Front* (Novato, Cal.: Presidio, 1994); and Dennis Showalter, "Mass Warfare and the Impact of Technology," in *Great War, Total War: Combat and Mobilization on the Western Front, 1914–1918*, ed. Roger Chickering and Stig Förster (New York: Cambridge University Press, 2000), 73–94.

30. Machine guns had already been invented in the 1880s, but they did not come into widespread use until World War I. The first flamethrowers arrived in late 1914, tanks in 1916. Paddy Griffith, *Battle Tactics on the Western Front: The British Army's Art of Attack, 1916–1918* (New Haven, Conn.: Yale University Press, 1994), 162; and John Frederick Charles Fuller, *Tanks in the Great War* (London: E. P. Dutton, 1920).

31. Cornelia Vismann, "Starting from Scratch: Concepts of Order in No Man's Land," in

War, Violence, and the Modern Condition, ed. Bernd Hüppauf (Berlin: Walter de Gruyter, 1997), 53.

32. Griffith, *Battle Tactics on the Western Front*, 136; Jürgens-Kirchhoff, "Verbrannte Erde," 786.

33. Hüppauf, "Räume der Destruktion," 108 and 113.

34. In August 1914, Germany had 232, France 165, and Britain 63 planes. Sönke Neitzel, *Blut und Eisen: Deutschland im Ersten Weltkrieg* (Zurich: Pendo, 2003), 124. See also Christoph Asendorf, *Super Constellation: Flugzeuge und Raumrevolution* (Vienna: Springer, 1997); Claude Carlier, "L'émergence d'une arme nouvelle: L'aéronautique 1914–1916," in *La Bataille de Verdun: Actes du colloque international*, ed. Claude Carlier and Guy Pedroncini (Paris: Economica, 1997), 37–52.

35. Hüppauf, "Räume der Destruktion," 117.

36. Robert Michaels, *Briefe eines Hauptmanns an seinen Sohn* (Berlin: S. Fischer, 1916), 69.

37. The British had first experimented with khaki uniforms in India in the 1850s. Martin Warnke, *Politische Landschaft: Zur Kunstgeschichte der Natur* (Munich: Hanser, 1992), 73.

38. Hüppauf, "Räume der Destruktion," 118; and Vismann, "Starting from Scratch," 55.

39. Jürgens-Kirchhoff, "Verbrannte Erde," 20.

40. Franz Marc, "Im Fegefeuer des Krieges," *Der Sturm* 7, no. 1 (1916): 1–12, esp. 1. See also Georges Lafond, *Covered with Mud and Glory: A Machine Gun Company in Action* (Boston: Small Maynard, 1918), 50.

41. "Der Kanonendonner ist häufig derart lebhaft, daß man keinen einzelnen Kanonenschuß hört, sondern nur ein stundenlanges ununterbrochenes Rollen." Letter from Friedrich Langhorst, dated January 2, 1915, quoted here from Ulrich and Ziemann, *Frontalltag*, 89.

42. Leed, *No Man's Land*, 124.

43. "Mit den Geräuschen des Krieges noch unvertraut, war ich nicht imstande das Pfeifen und Zischen, das Knallen der eigenen Geschütze und das reißende Krachen der in immer kürzeren Pausen einschlagenden feindlichen Granaten zu entwirren." Ernst Jünger, *In Stahlgewittern*, in *Sämtliche Werke*, 18 vols. (Stuttgart: Klett-Cotta, 1978–83), 1:32.

44. "[M]ir aus all dem ein Bild zu machen," ibid.

45. Habeck, "Die Technik im Ersten Weltkrieg—von unten gesehen," 108.

46. Wilfred Brenton Kerr, *"Shrieks and Crashes": Being Memories of Canada's Corps* (Toronto: Hunter Rose, 1917), 131.

47. "Der Schlachtendonner war fürchterlich geworden, daß keiner mehr bei klarem Verstand war." Jünger, *In Stahlgewittern*, 241.

48. Bloch, *Memoirs of War*, 96.

49. "Über den Ruinen lag, wie über allen gefährlichen Zonen dieses Gebietes, ein dicker Leichengeruch, denn das Feuer war so stark, daß sich um die gefallenen niemand kümmerte. Man rannte durchaus auf Leben und Tod, und als ich diesen Dunst im Laufen verspürte, war ich kaum überrascht—er gehörte zum Ort. Übrigens war dieser schwere und süßliche Hauch nicht lediglich widerwärtig; er rief darüber hinaus, eng mit den stechenden Nebeln des Sprengstoffs vermischt, eine fast hellseherische Erregung hervor, wie sie nur die höchste Nähe des Todes zu erzeugen vermag." Jünger, *In Stahlgewittern*, 100.

50. "Der Leichengeruch auf viele Kilometer im Umkreis ist das Entsetzlichste. Ich kann

ihn weniger vetragen als tote Menschen und Pferde sehen." Letter dated September 6, 1914, quoted here from Franz Marc, *Briefe aus dem Feld,* ed. Klaus Lankheit and Uwe Steffen (Munich: Piper, 1982), 8.

51. Quoted in John Ellis, *Eye-Deep in Hell* (New York: Pantheon, 1976), 59.

52. On the phenomenon of shell shock, see Paul Lerner, *Hysterical Men: War, Psychiatry, and the Politics of Trauma in Germany, 1890–1930* (Ithaca, N.Y.: Cornell University Press, 2003).

53. Eksteins, *Rites of Spring,* 142; and Hüppauf, "Räume der Destruktion," 117.

54. Bernd Hüppauf, "Kriegsfotografie," in Michalka, ed., *Der Erste Weltkrieg: Wirkung, Wahrnehmung, Analyse,* 875–92.

55. Bloch, *Memoirs of War,* 125.

56. "Alle Bäume zerschossen, die ganze Erde metertief zerwühlt von schwersten Geschossen, und dann wieder Tierleichen und zerschossene Häuser und Kirchen, nichts, nichts auch nur annährend brauchbar! Und jede Truppe, die zur Unterstützung vorgeht, muß kilometerweit durch dieses Chaos hindurch, durch Leichengestank und durch das riesige Massengrab." Letter dated November 5, 1914, quoted here from Philipp Witkopp, ed., *Kriegsbriefe gefallener Studenten* (Munich: Georg Müller, 1928).

57. A. A. Dickson, "Varieties of Trench Life," in *True World War I Stories: Sixty Personal Narratives of the War* (London: Robinson, 1997), 186.

58. W. Walker, "The Battle of Loos," in *True World War I Stories,* 39.

59. Frank Watson, "A Territorial in the Salient," in *True World War I Stories,* 26.

60. "Der Erste Weltkrieg lieferte die erste Anschauung von totaler Destruktion von Landschaft und ein Beispiel für unbegrenzte Herrschaft von Technologie über Natur und Raum." Hüppauf, "Räume der Destruktion," 117.

61. Entry dated January 19, 1915, in Roger Campana, *Les enfants de la "Grande Revanche": Carnet de route d'un Saint-Cyrien, 1914–1918* (Paris, 1920), 69; quoted here from Eksteins, *Rites of Spring,* 148.

62. "Vor Ungeziefern und Nässe und Kälte konnten wir nicht schlafen, stürzten ans Geschütz und lösten die todmüden Kanoniere ab, nur dass wir warm wurden." Letter from Christian Krull dated July 29, 1916, quoted here from Ulrich and Ziemann, eds., *Frontalltag im Ersten Weltkrieg,* 92.

63. *L'Argonnaute,* June 1, 1916, in Stéphane Audoin-Rouzeau, *Men at War, 1914–1918: National Sentiment and Trench Journalism in France during the First World War,* trans. Helen McPhail (Providence and Oxford: Berg, 1992), 38.

64. Bloch, *Memoirs of War,* 102.

65. "[H]eute ist ein abscheulicher Regentag; es ist alles so verschleiert, daß an Schießen nicht zu denken ist." Marc, *Briefe aus dem Feld,* 45.

66. The Battle of Verdun was supposed to have started on February 10, 1916, but did not commence until February 19. On the role of the weather during the battle's first weeks, see, among others, Alain Denizot, "Combattre à Verdun," in Carlier and Pedroncini, eds., *La Bataille de Verdun,* 33; Neitzel, *Blut und Eisen,* 85; and Robin Neillands, *Attrition: The Great War on the Western Front, 1916* (London: Robson, 2001), 77.

67. A. A. Dickson, "Varieties of Trench Life," in *True World War I Stories,* 186.

68. Bloch, *Memoirs of War,* 152.

69. Frank Watson, "A Territorial in the Salient," in *True World War I Stories,* 27.

70. Lafond, *Covered with Mud and Glory,* 221.

71. W. Walker, "The Battle of Loos," *True World War I Stories,* 39.

72. Smith, Audoin-Rouzeau, and Becker, *France and the Great War,* 89.

73. *Le Bochophage,* March 26, 1917, quoted here from Audoin-Rouzeau, *Men at War,* 38.

74. *Le Pépère,* April 21, 1916, quoted here from ibid., 42.

75. Ashworth, *Trench Warfare,* 5.

76. Maze, *Frenchman in Khaki,* 242, cited in Habeck, "Die Technik im Ersten Weltkrieg— von unten gesehen," 117.

77. Letter by Lt. Col. Sir Robert Tolerton to his fiancée on July 10, 1916, reprinted in Richter, *Chemical Soldiers,* 137.

78. Edmund Russell, *War and Nature: Fighting Humans and Insects with Chemicals from World War I to Silent Spring* (New York: Cambridge University Press, 2001), 26; and Paul Weindling, "The First World War and the Campaigns against Lice: Comparing British and German Sanitary Measures," in *Die Medizin und der Erste Weltkrieg,* ed. Wolfgang Eckart and Christoph Gradmann (Pfaffenweiler: Centaurus, 1996), 227.

79. Audoin-Rouzeau and Becker, *France and the Great War,* 89.

80. In 1916, Friedrich Prinzing published his internationally acclaimed work *Epidemics Resulting from Wars* (Oxford: Clarendon, 1916), in which he outlined some of the major epidemics that threatened front-line soldiers.

81. *Notes for Infantry Officers on Trench Warfare,* 53.

82. Major J. Graham, "Trenchfever," *Lancet* (September 25, 1915).

83. Gerhard Hirschfeld, Gerd Krumeich, and Irina Renz, eds. *Enzyklopädie Erster Weltkrieg* (Paderborn: Schöningh, 2004), 460. See also Jürgen Müller, "Die Spanische Influenza 1918/19: Einflüsse des Ersten Weltkrieges auf Ausbreitung, Krankheitsverlauf und Perzeption einer Pandemie," in Eckart and Gradmann, eds., *Die Medizin und der Erste Weltkrieg,* 321–42.

84. Gas had first been used by the French police in 1912. See Chrisoph Gradmann, "'Vornehmlich beängstigend': Medizin, Gesundheit und chemische Kriegführung im deutschen Herr 1914–1918," in Eckart and Gradmann, eds., *Die Medizin und der Erste Weltkrieg,* 131–54; Rolf-Dieter Müller, "Total War as a Result of New Weapons? The Use of Chemical Agents in World War I," in Chickering and Förster, eds., *Great War, Total War: Combat and Mobilization on the Western Front, 1914–1918,* 95–111. On the use of poison gas in World War I, see also Ludwig Fritz Haber, *The Poisonous Cloud: Chemical Warfare in the First World War* (Oxford: Clarendon Press, 1986); Olivier Lepick, *La Grande Guerre chimique: 1914–1918* (Paris: Presses universitaires de France, 1998); Dieter Martinetz, *Der Gaskrieg 1914/18: Entwicklung, Herstellung und Einsatz chemischer Kampfstoffe, das Zusammenwirken von militärischer Führung, Wissenschaft und Industrie* (Bonn: Bernard and Graefe, 1996); Donald Richter, *Chemical Soldiers: British Gas Warfare in World War I* (Lawrence: University Press of Kansas, 1992); and Ulrich Trumpener, "The Road to Ypres: The Beginnings of Gas Warfare in World War I," *Journal of Modern History* 47 (1975): 460–80.

85. Four types of gases were used in World War I: lacrimators or "tear gases" such as benzyl bromide; sternutators or vomiting gases; lung irritants such as chlorine, phosgene and carbon oxychloride; and skin irritants such as mustard gas.

86. Neillands, *Attrition,* 98; and Richter, *Chemical Soldiers,* 228.

87. Müller, "Total War as a Result of New Weapons," 101.

88. Audoin-Rouzeau, *Men at War*, 72.
89. Lt. Col. G. W. G. Hughes in April 1915, cited in Richter, *Chemical Soldiers*, 6.
90. Audoin-Rouzeau, *Men at War*, 72.
91. "Ein großer Teil aller Pflanzen war verwelkt, Schnecken und Maulwürfe lagen tot umher, und den in Monchy untergebrachten Pferden der Meldereiter lief das Wasser aus Maul und Augen." Jünger, *In Stahlgewittern*, 90.
92. Ellis and Cox, *The World War I Data Book*, 269–70. See also Denis Winter, *Death's Men: Soldiers of the Great War* (London: Viking, 1978), 251–54.
93. Michael Salewski, *Der Erste Weltkrieg* (Paderborn: Schöningh, 2003), 207.
94. John Jeanneney, "The Impact of World War I on French Timber Resources," *Journal of Forest History* 22 no. 4(October 1978): 226–27, esp. 226.
95. "Die schönen Dörfer waren erst Ruinen geworden, dann Trümmerhaufen, schließlich Ziegelstätten; die Wälder erst Lücken und Knäuel, dann Leichenfelder bleicher Stümpfe, schließlich Wüste." Arnold Zweig, *Erziehung vor Verdun* (Leipzig: P. Reclam, 1971), 101.
96. Mosse, *Fallen Soldiers*, 109.
97. "Ein Wald ist ein Volk. Ein zerschossener Wald ist ein gemeucheltes Volk." Ernst Toller, *Eine Jugend in Deutschland*, vol. 4 of Ernst Toller, *Gesammelte Werke*, ed. John M. Spalek and Wolfgang Frühwald, 5 vols. (Munich: Hanser, 1978), 64.
98. "Das Schicksal dieses Waldes ist mit dem meinigen aufs innigste verquickt und verworben. Nicht nur Kamerad ist mir der Wald geworden, sondern auch starker Schutz; ein Schild, der das gegen mich geschleuderte tödliche Blei und Eisen abwendet . . . sich ins Herz bohren läßt; sich selbst wehrlos dem Tode beugt, auf das ich lebe. Oft wenn dicke Wolken die Sonne verdrängt haben . . . der Tag erfüllt ist von Wehmut und Trauer, dann perlen die hellen Tropfen aus den zerschundenen Kronen, wie Tränen eines unendlichen Schmerzes." H. O. Rehlke, "Der gemordete Wald," in *Die Feldgraue Illustrierte: Kriegszeitschrift der 50. Infantrie Division* (June 1916), 12.

CHAPTER 5

TOTAL WAR?

Administering Germany's Environment in Two World Wars

Frank Uekötter

On February 2, 1943, the Sixth Army of the German *Wehrmacht* surrendered in Stalingrad. Two weeks later Joseph Goebbels declared "total war" in his infamous Sportpalast speech, and two weeks after that a group of bureaucrats met in Berlin to discuss what seemed a most untimely topic: the conservation of the Wutach Gorge in southwest Germany. The Schluchseewerk power company intended to use water from the Wutach River to produce electricity during hours of peak demand. Only by diverting water from the Wutach, the planners argued, would the Schluchseewerk project be efficient and complete. The Wutach was not, however, just a scenic river flowing through a beautiful gorge carved over the millennia; it was also a nature reserve. In 1939 a ministerial decree had granted protection to an area of 579 hectares (1,430 acres) along the banks of the Wutach, a sizable area by German conservation standards.[1] The Schluchseewerk plan involved building a dam at the confluence of the Haslach and Gutach rivers, which join to form the Wutach, effectively submerging the upper stretch of the nature reserve. To make matters even worse, the plan called for diversion of the lion's share of the Wutach's water to the company's power plants, leaving a much-reduced stream in the gorge. Conservationists recognized that such drastic changes would completely transform one of southern Germany's most scenic natural environments.[2]

Paragraph 20 of the National Conservation Law of 1935 required that every proposed alteration of the landscape be approved by the conservation administration.[3] However, conservationists trying to enforce this requirement were fighting an uphill battle. In January 1938 a ministerial decree cited numerous cases in which conservation officials had been consulted either not at all or so late in the planning process as to prevent any meaningful participation.[4] It is thus hardly surprising that Hermann Schurhammer, the chief conservation advisor in the state of Baden, wrote in a letter of December 21, 1941, "that there is reason to fear that construction will simply ignore some of the objections raised" in his reports.[5] But because Baden's minister of education and cultural affairs (Badisches Ministerium des Kultus und Unterrichts als Höhere Naturschutzbehörde) had earlier expressed his unqualified support for Schurhammer's stance, Schurhammer's objections could not be ignored altogether.[6] This is all the more remarkable since Schurhammer did not simply recommend some design changes but opposed the construction of a dam in the Wutach Gorge on principle. For Schurhammer, the gorge was "the most valuable nature reserve in the state of Baden" and "not one of those things that one can evaluate adequately in terms of money, calories, or kilowatt hours."[7]

As a result, the project was effectively stalled. The Inspector General for Water and Energy (Generalinspektor für Wasser und Energie) had given priority to the project in September 1941, calling for an immediate start to construction. The decree also mandated that the Schluchseewerk adhere to the common licensing procedures, however, which led to prolonged negotiations between the two factions.[8] In March 1942, Schurhammer reaffirmed his opposition to the project, arguing that the dam and the diversion of water would make the Wutach Gorge "worthless as a nature reserve."[9] Nevertheless, the Schluchseewerk was still hoping for a decision that would "do justice to both energy production and the protection of nature," while the Inspector General enlisted the support of Alwin Seifert, the official German Landscape Advocate (Reichslandschaftsanwalt) of the Inspector General for German Highways (Generalinspektors für das deutsche Straßenwesen) and one of the most powerful conservationists of the Nazi era. After allegedly traveling through the reserve for a mere ninety minutes, Seifert found no reason to disallow the project so long as the dam was designed in an aesthetically pleasing way.[10] The conflict continued to escalate, and with the local administration being unable to reconcile the divergent views, the matter went to Berlin for arbitration. In the end, it was up to the Reich's Forest Service (Reichsforstmeister) as the final authority on conservation issues to make a decision. The Forest Service invited all parties to the aforementioned meeting on March 3, 1943, where Schurhammer, supported by an official from the Ministry of Education and Cultural Affairs, reiterated his fundamental opposition despite the wartime demand for energy, while representatives of the Schluchseewerk and the inspector general stressed the Wutach's suitability for electric power production.[11] In the end, the Forest Service reluctantly granted an exception, observing that recent revisions of the plans had "facilitated" its decision.[12] Under pressure from the conservation administration, the Schluchseewerk had made significant concessions: as late as September 1942, it had been planning to release only 17 percent of the river's total flow into the original bed of the Wutach.[13] By March 1943 that figure had doubled to 35.3 percent—not enough to appease the conservationists but a remarkable concession nonetheless given the condition of "total war."[14]

Obviously, this episode does not fit neatly into common interpretative schemes of Nazi (National Socialist German Workers' Party [NSDAP]) rule during World War II, especially when one considers that the conflict over the Wutach Gorge was not due to Fritz Todt's penchant for nature protection.[15] The conflict continued long after Albert Speer took charge of the war economy, a development that many historians of World War II have linked to the unprecedented intensification of war production since 1942.[16] It is even more remarkable that the energy utility and its allies within the war administration never dared to challenge the legitimacy of the conservationists' issue. At one point, Robert Wagner, the administrative head in the state of Baden (Reichsstatthalter) and a staunch supporter of the Schluchseewerk project, questioned whether the conservationists were "sufficiently taking into account the precarious situation of our electric power supply,"[17] but that is as far as it went. The Reich's engineers never attempted to bypass the conservationists; neither did they seek to start construction before a ruling was made on the conservationists' objections. Clearly, the story of the Wutach Gorge demonstrates that even in a period of "total war," environmental issues did not disappear altogether from the German political agenda. Except for the last few months of World War II, the environment remained a legitimate concern; and although war conditions were hardly conducive to effective conservation, there was at least some effort made to preserve the country's scenic landscapes and natural resources.

In a recent publication on German conservation in times of war, Willi Oberkrome focuses on the impact of wartime conditions on conservation ideology. Employing martial terminology, Oberkrome spoke of "core troops" in "times of crisis," suggesting that conservation ideology became increasingly radical during the two world wars.[18] It is, however, difficult to understand the conflict over the Wutach Gorge by means of an interpretation that puts conservationist theory center stage. Even Schurhammer's last report on the Wutach Gorge, written when he was under pressure to mobilize every possible argument in favor of his stance, was remarkably free of ideology. Instead he stressed the biological and geological peculiarities of the Wutach Gorge.[19] This is somewhat astounding in view of the fact that conservationists in the 1920s had asserted that protecting the Wutach Gorge would be an act of patriotism;[20] and the well-known work of German landscape planners in Eastern Europe provides shocking proof of the extent to which conservation issues could be charged with Nazi ideology.[21] Obviously, the radicalization of conservation ideology was a far more ambiguous and partial process than some scholars have suggested.[22]

On a more fundamental level, Oberkrome's reading fails to take into account the institutional background of conflicts over conservation issues during the world wars.[23] In the dispute over the Wutach Gorge, the key players held places within the administration, and their word carried weight only because they were part of a bureaucratic network. Of course, Schurhammer and others would routinely invoke public opinion to bolster their claims, but popular support was rarely manifested in resolutions by groups or associations that shared their view, let alone in open protest. It is therefore imperative to see environmental conflicts during the two world wars as *administrative* conflicts. The Wutach project stalled only because the conservation administration refused to approve it; and this administrative forum was the main—perhaps the only—one for discussing environmental problems in these periods. In fact, there is room to argue that administration was indeed the crucial aspect of the entire Nazi history of conservation; while the ideological rapprochement remained ambiguous at best, the conservation community made great strides in conservation practice, making Nazi Germany the only European country with a boom of conservation work in the 1930s.[24]

Administering the Environment during World War I

When war broke out in August 1914, the European powers expected the conflict to be brief. In spite of increased military budgets in the years before World War I, no European country possessed an economy that could sustain a prolonged war, and most did not even have a blueprint for developing such an economy. Therefore the German army, like its counterparts, quickly ran short of ammunition, and the urgent need to supply the most basic tool of modern warfare posed an unprecedented challenge to the German economy. How it responded to that challenge and the social repercussions of that response constitute a well-developed field of historical study.[25]

It is understandable that under these conditions interest in addressing environmental concerns was low. For example, an official of a war production plant in Erfurt, after being reprimanded for making trial runs with engines at night, noted that "If [the complainant] or one of his sons were under fire at Verdun, they would long for the sound of one of our engines."[26] A public health officer readily agreed, not only attesting that the noise was not detrimental to human health but also offering the opinion that the complainant "should thank God that he doesn't have to endure the nuisances that our enemies are causing in this war."[27] When further complaints were made about smoke from a steam-powered incline in the city of Stuttgart in 1915, officials played for time; two of them also vented their anger by speculating in internal communications about the "ideal" procedures for handling such complaints. The proposal they came up with was telling: they would require each complainant to place a deposit that would be forfeited to the treasury if administrative investigators ultimately dismissed the complaint.[28] When war broke out, committees for smoke abatement in Munich and Chemnitz suspended their work, never to resume it even after the war ended.[29] Environmental concerns clearly ranked near the bottom of any list of wartime priorities, and a 1918 legal brief for a debate in the upper house (the Bundesrat) of Germany's parliament conceded that some wartime installations "failed to meet standards for the protection of workers and the environs to such an extent that their speedy closure is desirable."[30]

Yet it is interesting to note that environmental issues remained on the agenda to a limited extent. Raising questions about environmental problems was by no means taboo. In 1915 the Prussian Ministry of Education complained of smoke from a nearby hotel, while the mayor of Lucherberg, a town in the lignite mining district west of Cologne, objected to the expansion of a plant for the production of briquettes.[31] Officials processed complaints just as they had during peacetime, collecting reports from technical and medical experts before rendering a decision; bureaucrats were more lenient in their calls for abatement and granted more time for repairs, but they did not simply ignore existing rules and regulations. The administration therefore followed "business-as-usual" procedures even when considering whether to license a war production plant. In November 1916 authorities in Bremen deliberated for more than a month before approving the construction of a foundry for defense purposes.[32] The gasworks of Bremen filed its plans for a new coke oven on July 24, 1917, and received approval from the administration a month and a half later.[33] Even engineers who in 1917 asked for "expeditious treatment" of their request for permission to build a submarine-fuel plant had to wait more than two months so that complaints from local residents could be addressed in an orderly fashion.[34] German officials clearly wished to adhere to the letter of the law and to follow prewar administrative routine.

Some of their proceedings certainly bordered on the farcical. In 1916 a botanical journal published an article on forest death as a result of artillery fumes, written by a soldier who had observed this phenomenon in a park in St. Mihiel, a French town on the Meuse River south of Verdun, and his findings even became the subject of lively academic debate.[35] In a few cases the administration took steps to curb pollution. In 1915, for example, the factory inspector of Erfurt required a smoking bakery to take several corrective measures, with the only acknowledgment that the country was at war being the allowance of "sufficient time."[36] In 1916 a mining company in the Ruhr Valley filed plans with the local administration for the construction of a fly-ash filter to reduce noxious emissions.[37] The mayor of a small town near Duisburg persuaded a branch of the Krupp industrial empire to remedy a dust problem in 1915.[38] The mayor of Duisburg secured the installation of a dust filter in a plant that produced graphite, and even brought about the temporary closure of a plant in 1916 after a number of influential residents complained of its awful smell.[39] Despite the exigencies of war, there remained a conviction that the administration was obliged to act on any problem that qualified as a genuine hazard or nuisance. As Konrad Wilhelm Jurisch wrote in 1916, "Even though one tolerates many disadvantages during the war without much ado, there is no need to increase them through negligence."[40]

One hallmark of the German approach to pollution control has always been its inconsistency. Whereas the French law of 1810 defined three classes of industry with different standards and procedures, the German administration favored a case-by-case method.[41] Surviving files display an astonishing variety of administrative behaviors, ranging from lethargic decrees that allowed excessive amounts of pollution to strict measures and even the closing of factories.[42] This administrative chaos was (and is) a significant burden for conservationists in Germany, but from the viewpoint of the wartime bureaucracy, it offered an important advantage in that it gave considerable latitude to officials in dealing with pollution problems. Instead of changing or suspending formal rules and regulations—an inherently cumbersome process, as every expert on administration knows—they could simply adopt a more lenient stance or choose to accept high levels of pollution for the time being. In an observation typical of the time, Bremen's factory inspector during the war years noted in his report on two factories producing for the military that their deleterious impact on the neighborhood would have to be tackled "in calmer times."[43] If officials were authorized to make this kind of decision in their own right, there was obviously no need for a formal suspension of antipollution rules.

If the administration of environmental law nonetheless became a topic of debate, this was due not to disagreement about the goal of pollution control but to procedural concerns. The German trading regulations (Reichsgewerbeordnung) required a special licensing procedure for a list of plants that presented particular environmental hazards, and paragraph 17 of the trading regulations required that all construction projects for these types of plants be open for public review for four weeks (an interval later reduced to two weeks) in order to give citizens a chance to file objections.[44] For decades this procedure had been generally uncontested, but in wartime, bureaucrats suddenly realized, it created a significant problem: allowing public access to plans for a munitions factory, for example, invited espionage. Even routine announcements of new construction projects in the newspapers became impossible. At the same time, urging Parliament (the Reichstag) to change the law was out of the question. Not only was it time-consuming, it also clashed with the general tendency of the German government to marginalize the Reichstag during the war and to prevent any increase of parliamentary power. In other words, the challenge for the bureaucrats was to somehow allow the construction of new factories without openly breaking the law. The solutions they found show that German bureaucrats are not necessarily unimaginative people.

Kurt von Rohrscheidt's legal brief on the trading regulations of 1901 noted that the law was targeting installations built "for a certain amount of time."[45] But what constituted a "permanent" business? Obviously impressed by its own propaganda about the impending victory, the Bavarian Ministry of War ruled that installations built exclusively for war production purposes, since they would exist for such a short period, did not qualify as "permanent"; in other words, war production in Bavaria was legally "temporary."[46] Administrators in the state of Württemberg found a second loophole that was even more daring: while acknowledging that war production plants fell under the provisions of the trading law, and that failing to apply for a license constituted a breach of the law, they contended that trading regulations did not explicitly require *prosecution* of violators. Thus war production in the state of Württemberg was technically illegal, but the government chose not to enforce the law.⁴⁷ The Prussian Ministry of War had used a similar tactic in 1915 when it instructed an official in the Anhalt region to formulate conditions for construction as in peacetime but to postpone the official inauguration of the licensing process "until the war is over."⁴⁸ Other Prussian officers found yet another loophole in the state-of-siege law (Belagerungszustandsgesetz), which authorized the military to license war production plants. This law was valid only while the nation was at war, and this created an unexpected problem when the Reich seemed to be on the verge of victory in the spring of 1918: if the war were to end suddenly, the entire German economy would instantly become illegal—a scenario that must have been a terrible nightmare for any Prussian bureaucrat who lived up to his stereotype.⁴⁹ Months of negotiations in Parliament's upper house finally led to an acceptable solution in October 1918.⁵⁰ As the German army retreated, the kaiser's authority disintegrated, and the Reichstag paved the way for democracy, the Imperial German government was engaged in validating licenses for war production plants.⁵¹

These ghostly debates over wartime licenses never became known to a broader public, and the administration's legal makeshifts were never tested in court. The procedural troubles returned in the 1930s, however, when the Nazi government was taking the first steps toward German rearmament. How could officials license a secret plant without breaking the law? Unlike administrators during World War I, the Nazis did not have to worry about parliamentary obstacles to new legislation. Instead they simply inserted a new paragraph (numbered 22a) into the trading regulations. In the summer of 1934, the administration began using this paragraph to legitimize a secret licensing procedure, with the key requirement being that the project had to be "in the public interest."⁵² If further clarification was needed, the German Ministry of Trade and Commerce (Reichswirtschaftsministerium) made any interpretative ambiguities obsolete with a decree of October 30, 1934, which stated that this condition could be presumed as a given when "dealing with plants that produce military supplies."⁵³ The ministry supplemented this regulation in 1941 by granting licenses for a defined time period in cases where an open-ended license was unacceptable.⁵⁴

Finally, it is noteworthy that there was no attempt during World War I to use the conflict as an excuse to overturn existing environmental legislation. The state and industry clashed over countless issues as the war economy developed, but environmental regulation never became controversial in this context. As I have argued elsewhere, industrialists had largely accepted the government's right to intervene against excessive pollution long before the turn of the century, and both bureaucrats and businessmen adhered to this line of reasoning after war was declared, making environmental regulation, strange as it may sound, an island of peace in a sea of controversy.⁵⁵ In November 1917, Samuel Insull, president of the Chicago Edison Company, which had been repeatedly fined for smoke pollution, used his position as chairman of the Illinois

State Council of Defense to call for suspension of the local smoke ordinance.⁵⁶ In Germany, extant files show no traces of similar gambits; administering the environment was simply a matter of bureaucratic routine.

Administering the Environment during World War II

In early 1943 the SS began to use the large crematoriums II–V at the Auschwitz-Birkenau extermination camp.⁵⁷ At about the same time, the camp's building department (Zentral-Bauleitung der Waffen-SS und Polizei Auschwitz) was negotiating with the trade administration in Bielitz in Upper Silesia over the construction of a "heating plant at the KL-Auschwitz." We know this because the trade administration requested a report on the maximum fly-ash output to be written by the Federal Agency for Water and Air Quality (Reichsanstalt für Wasser- und Luftgüte), Germany's highest scientific authority on pollution problems. The agency reacted just as it had to previous requests: it asked for a map of the environs—presumably the reason why the report was never written.⁵⁸ Even to a regime well known for both cynicism and administrative chaos, the building department's request was deeply troubling. The disturbing parallelism of horrendous crimes and bureaucratic routine has long been recognized as a hallmark of the Nazi regime.⁵⁹

One side effect of the request was to show that the Federal Agency for Water and Air Quality was still up and running far into the fourth year of the war. In fact, the agency's position was in some respects stronger than it had been in 1939. A decree of the German Ministry of Trade and Commerce of February 18, 1942, directed local authorities to consult with the institute's experts if they wished to license certain plants or if they had "doubts concerning the conditions necessary for the protection of the neighborhood."⁶⁰ The list appended to the decree specified large steam boilers; metallurgical factories; plants producing magnesium, carbide, or cement; and several types of chemical plants. Numerous requests for advice followed this decree.⁶¹ The agency was also elevated from a Prussian institution to an agency of the Reich in April 1942.⁶² The agency had addressed air-pollution problems outside Prussia before, but now its authority was officially recognized throughout Germany.⁶³ Thus the general picture of air-pollution control in the early 1940s looked surprisingly similar to that of World War I: business as usual, just under more difficult conditions. In the spring of 1940 the NSDAP's district organization (Kreisleitung) protested the construction of a new foundry in Leipzig that was part of an urgent rearmament plan.⁶⁴ And in Bielefeld, city administrators cited air-pollution concerns when they refused a printer's request to dismantle his shop's smokestack because "during the last two air-raids on Bielefeld, enemy pilots [had] regularly used [his] chimney as a target when they dropped bombs in the environs of [his] factory."⁶⁵

The conservation administration's files ratify this impression. During World War I, proceedings on nature protection issues were rare. In 1917 the city of Bitterfeld contacted the Prussian Agency for the Preservation of Natural Monuments because they feared for a nearby forest; and in response to a request from the Ukrainian government for information, the Prussian Agency readily supplied a report on conservation in August 1918, followed by a draft for a Ukrainian national conservation law in November.[66] In general, however, conservation authorities of the 1910s typically focused on such things as outstanding trees, cliffs, or other small-scale "natural monuments" that did not stand in the path of a war economy. If the situation was different in World War II, this was due to a gradual expansion of the definition of conservation in the interwar years. Conservationists moved beyond small nature reserves to include beautiful landscapes and ultimately the entire German countryside in the realm of protected phenomena.[67] Paragraph 20 of the national conservation law of 1935 required approval from the conservation administration for any project that would cause major changes to the landscape.[68] That paragraph made conflicts with the war economy inevitable—conflicts showing that conservation concerns were shared by a wide array of institutions. In a dispute over a proposed coal mine near the scenic Porta Westfalica (literally the "door to Westphalia," the point at which the Weser River descends from the Weserbergland mountains to the lowlands of northern Germany), the protest of the conservation administration was echoed by both district and regional administrations; by tourist associations of the district, the region, and the Reich; and even by the Prussian Minister of Science (Minister für Wissenschaft, Kunst, und Volksbildung).[69]

But the continuation of prewar activities (if on a somewhat smaller scale) did not last through the entire war. On July 1, 1943, the conservation administration was ordered to stop all routine work and to restrict its activity "to the extent indispensable for the war."[70] Air raids in November 1943 and January 1944 destroyed the offices of the German Bureau for Nature Protection (Reichsstelle für Naturschutz). Luckily, Hans Klose, director of the bureau since 1939, had already transferred a good part of the bureau's files to the town of Bellinchen, which became the impromptu seat of the Bureau for Nature Protection until the advancing Red Army forced it to be moved again in early 1945. Klose relocated his bureau to Egestorf, a small town some thirty kilometers south of Hamburg in rural Lower Saxony, where he waited out the rest of the war.[71]

The situation of the Federal Agency for Water and Air Quality during the last months of the war was somewhat luckier. A letter of January 1944 reports significant damage from air raids, but also notes that the agency was already operating normally again for the most part.[72] The agency never left Berlin, and as the rich archive attests, damage from aerial bombing was limited; the Agency for Water and Air Quality is one of few German institutions whose files survived the war almost intact. Although short on personnel, the agency issued thirty-six reports on pollution problems in 1943 and twenty-two reports in 1944.[73] The agency was evidently still operational as late as March

Total War? 101

1945, when the author of an internal memo recommended that a document be reviewed again "after the war's end."⁷⁴ However, it seems that the last two war years saw little more than token efforts, especially compared with the activity of previous years. The final meeting of the committee on dust technology of the German society of engineers (VDI-Fachausschuß für Staubtechnik), the most important committee of experts in the field of air pollution, took place in October 1943.⁷⁵

Of course, administering the environment during World War II was an uphill job, and not only due to reductions in personnel. While officials did succeed in mitigating a number of environmental problems, the war deprived them of the necessary means to do so in many cases.⁷⁶ In some respects, this lack was actually a by-product of technological progress. Dust filters, which were rudimentary devices during World War I, had by 1939 become high-tech machines that required time, skill, and costly materials to build, all of which were increasingly scarce. A long-standing conflict over fly-ash emissions from the Herdecke power plant in the Ruhr Basin provides a case in point. In the late 1930s, officials finally convinced plant operators to retrofit the boilers at Herdecke with efficient fly-ash collectors, but in 1940 the work was suspended because the war administration refused to supply iron for an electrostatic precipitator.⁷⁷ When asked in December 1943 for advice on the construction of new boiler units in Upper Silesia, the Federal Agency for Water and Air Quality, knowing that iron for a precipitator or a cyclone was unavailable, recommended using simple ad hoc devices and restricting the cultivation of sensitive plants in the environs.⁷⁸ When half a dozen officials met in Lüdenscheid to discuss the impact of an aluminum smelter on local trees in 1944, they were unanimous in their critique of the damage but concluded that nothing could be done for the duration of the war.⁷⁹ One wonders why the meeting had been called at all.

The Upswing of Environmental Policy Debates during World War II

In March 1943, Hans Klose proposed a thorough inventory of nature reserves in the Caucasus region.⁸⁰ Given the contemporary situation, it is hard to conceive of a more untimely request; the German army had never actually conquered the Caucasus, and after its hectic withdrawal from the region in the previous weeks, such an inventory could not have been considered urgent in any respect. Yet Klose's initiative must be seen in the context of the rivalries for authority over conservation matters in the occupied Eastern European territories.⁸¹ It reflected the general trend in conservation work during the war years, when a boom of conceptual debates on environmental issues occurred. Rather than addressing the critical concerns of the day, officials were ready to adopt a long-term perspective and to discuss questions that were of no immediate relevance. The best-known example is the infamous work of landscape planners in Eastern Europe.⁸² However, the vigor of debates in this field is somewhat obscured by the fact that the 1920s and '30s had been a boom time for landscape planning and landscape architecture, which made wartime discussions appear to be merely a continuation of prewar debates.[83]

The oddness of these debates becomes more apparent when one focuses on the area of air-pollution control, where controversies about more general strategies were virtually dormant at the start of the war. All of a sudden, officials were ready to concede that the same procedures they had followed and even defended in previous years were actually deficient. For example, a memorandum from the Department of the Interior (Reichsministerium des Innern) of August 1943 predicted that air-pollution control would become even more urgent "after the war" and cited a report of the Federal Agency for Water and Air Quality on the industrial region of Upper Silesia.[84] This report was noteworthy in itself; for the first time, the agency had conducted a systematic study of regional air pollution. In October 1940 the agency took air samples in a formerly Polish area of some hundred square kilometers in the center of the industrial district. Due to the complexity of the task, the agency did not complete its report until April 26, 1942. Still, the findings and methodology departed markedly from the agency's previous reports. First, the agency had abandoned its former case-by-case approach, in which recommendations were always focused on a specific polluter. In the Upper Silesia study the focus was on air quality in general, which represented a fundamentally new perspective; the agency concluded that "the present situation is intolerable from the viewpoint of public health."[85] Second, the agency adopted a holistic approach to the problem, recommending not only measures to curb emissions but also land-use changes and reforestation efforts.[86] Third, the agency investigated the problem of domestic smoke for the first time, suggesting "radical steps" in the cities of Kattowitz and Königshütte.[87] Fourth, the report hinted at a chronic acidification of the soil as a result of sulfur dioxide emissions, an issue that did not draw major attention in Germany until entire forests were threatened in the 1980s, a phenomenon known in Germany as "forest death."[88] Finally, the agency criticized the practice of offering monetary compensation for pollution damage, noting that the problem had to be cured at the source.[89] The last observation seems like a truism from today's perspective, but given the ubiquitous payments made both over and under the table in countless sulfur emission conflicts, it was an almost revolutionary demand. In 1944, an official from the federal agency noted that "the air in industrial areas is barely better nowadays than a generation ago."[90] Perhaps the war was instilling a new sense of integrity in people who had grown tired of proclaiming that things were under control.

Historians have found that postwar advances in a number of fields in fact derived from plans developed during the war years.[91] In the field of environmental policy, it seems that the impact on postwar reforms was very limited, if it existed at all. For example, the German Ministry of Trade and Commerce proposed to revise the list of plants that required a special license due to their

Total War? 103

deleterious impact, but the bureaucrats never came up with corresponding legislation.[92] Also, work was begun in 1944 on a new handbook for these plants that was to include up-to-date technical information and standards, but the resulting text was highly disappointing.[93] In the field of nature protection, demands for reform were scant, as the national conservation law of 1935 had fulfilled almost every wish of the conservation community.[94]

Conclusion: Motives and Driving Forces

Before "the environment" became a catchphrase implying an intrinsic link between conservation and air-pollution control, the two fields had little to do with each other. The protection of nature was supported by numerous civic groups that focused almost exclusively on the countryside; and conservation was usually housed in departments of education and culture until Hermann Göring made it part of his Forest Service in 1935.[95] In contrast, air-pollution control dealt with urban agglomerations and was closely linked with issues of workplace safety, making departments of labor its logical administrative home.[96] What both fields had in common was the fact that neither could claim to be instrumental to winning the war; occasionally, officials in charge of air-pollution control reported that emissions were destroying precious crops, but that fact never influenced military strategy or administrative policy in either of the world wars.[97] This underscores the need to look at the bureaucratic context, rather than the environmental problems themselves, in order to account for administrative continuity. To a large extent the persistence of pre-war procedures was due to the *bureaucratic* character of work in both fields: there were laws and regulations that had to be enforced, there were civil servants waiting for instructions, and there was an established routine carried over from the prewar years; officials simply tried to adhere to this routine as much as they could.

It is tempting to explain the surprising persistence of "business as usual" with Hegelian dictums about the state as "the realization of reason." But the connections between administrative activity in World War II and traditional notions of German statehood are not as clear-cut as one might expect them to be. In fact, Franz Neumann, in his wartime study of "the structure and practice of National Socialism," suggested a fundamental incompatibility between the rational administrative state and "National Socialist 'dynamism,'" remarking that "Hegel cannot be held responsible for the political theory of National Socialism."[98] Such an interpretation certainly overstates the actual situation, however, for bureaucratic routines usually have a momentum all their own that even a self-proclaimed "revolutionary" government like the Nazi regime can overcome only to a certain degree. In many wartime environmental conflicts it was not so much the activity of the environmental administrators but rather the readiness of other bureaucrats to listen to their demands, and to accept the legitimacy of their concerns even if they disagreed on the issue, that must in retrospect be seen as remarkable. And if giving conservationists a hearing meant that an important energy project was stalled for a year and a half, as in the case of the Wutach Gorge, that was simply the price that had to be paid.

Such a functionalist interpretation ignores the fact that officials had a personal interest in the continuation of their work. To put it simply, a desk far from the front lines was one of the safest places to be between 1939 and 1945. It therefore seems reasonable to assume that the intensification of policy debates during World War II was due at least in part to bureaucrats' efforts to ensure their own indispensability. Officials would never dare to mention this rationale in their correspondence, but in some cases the circumstances were revealing. For example, the decree of the German Ministry of Trade and Commerce of February 18, 1942, which promised more work for the Federal Agency for Water and Air Quality, was remarkably similar to a draft that the agency had sent to the ministry about a month earlier.[99] The timing of the ministry's proposal to revise the list of plants requiring a special license is also suggestive; it was issued on June 25, 1941, three days after the German attack on the Soviet Union.[100] Artificially generating paperwork was a strategy that relied on other conditions to succeed. Had the German war administration been more effective in transferring lower-echelon bureaucrats to the battlefield during World War II, this strategy would surely have failed. But with the Department of the Interior dragging out the completion of an inventory of administrative personnel until August of 1940, supervisors were able to resist recruitment by giving the impression that even with the current staff, their departments were terribly overworked.[101] Only toward the end of the war was this strategy finally reaching its limit.[102]

A third rationale involved the Nazis' subliminal fear of a loss of popularity. Even during the war, the Nazi regime carefully monitored and responded to public opinion; not until the last months of the war did the regime abandon consideration of initiatives that were unrelated to the war effort.[103] This is important because both air-pollution control and the protection of nature had long proven to be popular concerns, as countless newspaper articles and statements from civic groups in the first decades of the twentieth century demonstrate. This may explain the general neutrality of the Nazi leadership in environmental conflicts. No legislation intended to undermine restrictions on air pollution was ever passed or even proposed, and in the field of conservation such regulations were issued only toward the end of the war. In fact, a decree of July 1, 1943, acknowledged the general importance of the issue and even encouraged conservationists to intervene if they deemed such action necessary: "since wartime planning may lead to permanent damage to nature and the landscape, the conservation administration will need to vigorously represent the concerns of nature preservation and landscape protection in future cases."[104] This reluctance to take unpopular actions may ultimately have been a stronger motivating force for environmental protection during World War II than any ideological agenda. The absence of ideological overtones in Schurhammer's defense of the Wutach Gorge in 1942 is by no means exceptional; in his speech celebrating the transformation of the Prussian Institute for Water, Soil, and Air Hygiene into the Federal Agency for Water and Air Quality in April 1942, the Secretary of State for Public Health (Reichsgesundheitsführer) stressed the agency's forty-year tradition and hoped for its continuation, making no attempt to link its work with quintessential Nazi goals.[105] In his work on the landscape advocates (Landschaftsanwälte) associated with construction of the Autobahn in the 1930s, Thomas Zeller has argued that the prominence of ideology in their rhetoric corresponded to their lack of influence.[106] One might therefore interpret nonideological statements by conservationists in Nazi Germany as an indication of strength; conservationists were confident that by resorting to their legal rights, and by representing a highly popular idea, they could make headway even under the adverse conditions of World War II.

This brings up what is arguably the most difficult question to address — namely, whether the activity of the bureaucracy brought with it a commensurate effect for the protection of the environment. Our current knowledge about German conservation history includes little about what happened in the countryside during the war. To make matters worse, the war followed a boom period of conservation that began with the passage of the national conservation law of 1935, which resulted in a flurry of activities that Hans Klose would later celebrate as the "heydays of conservation."[107] To some extent, that boom was fading off during the first war years, making it a matter of perspective whether it was a decline of activity or a stunning persistence of conservation work under averse circumstances. Finally, one should generally be cautious in labeling conservation efforts as completely ineffective, and not only because conservationists themselves are usually the first to render this verdict.[108] At the very least, one should credit the administration with preventing an even greater environmental imbalance. Overall, the environmental toll in Nazi Germany could have been far worse, as a look at Stalinist Russia or Maoist China quickly shows.[109] Finally, it is important to include the unexpected byproducts of conservation in any historical assessment. In the case of the Wutach Gorge, the conservation administration had proved unable to prevent the licensing of the dam-building project; but because conservationists' objections had stalled the project for a year and a half, the war had progressed too far for construction to start. The Wutach is therefore still the wild river that it has always been, without dams or diversions.

That fact is also due to civic protest that prevented execution of the Wutach project during the 1950s, when the Schluchseewerk revived its wartime plans and tried to invoke the 1943 license. Bureaucratic inertia could impede a project for a while, but public protest over an entire decade gave a much different kind of force to the voice of the conservationists. Nor was protest by third parties completely absent from the Wutach conflict of the war years. In December 1942 the natural science department of Freiburg University passed a resolution against construction of the dam.[110] But this resolution does not bear any comparison with the deluge of protest that was seen in the 1950s, when no less than 185,000 citizens signed a petition for the preservation of the Wutach Gorge.[111] At the same time, the sustained protest of the 1950s expressed the limits of wartime normalcy; a wave of public protest on a conservation issue was unthinkable in the Nazi era and especially during the war. Despite the abnormal circumstances that prevailed, environmental practices remained remarkably stable—normal environmental protection, or an attempt at such, during abnormal times.

Notes

1. Verordnung über das Naturschutzgebiet Wutach-Gauchachtal, July 26, 1939, EA 3/102 No. 29, Hauptstaatsarchiv Stuttgart (hereafter HStAS).

2. In this essay, I use the terms *conservation* and *conservationist* as synonyms for nature protection and people concerned with it. *Conservation* is not meant as an antonym to *preservation*, as it is in American environmental history; and any allusion to U.S. concepts of "wise use" would be misleading.

3. See Karl Cornelius, *Das Reichsnaturschutzgesetz* (Bochum-Langendreer: Pöppinghaus, 1936), 37.

4. *Reichsministerialblatt der Forstverwaltung* 2 (1938): 43.

5. Schurhammer to Badischer Finanz- und Wirtschaftsminister, December 21, 1941, HStAS. All translations of quotations from German sources are my own.

6. Minister des Kultus und Unterrichts als Höhere Naturschutzbehörde to the Badischer Finanz- und Wirtschaftsminister, July 4, 1941, HStAS.

7. Gutachten der Landesnaturschutzstelle Baden, November 30, 1942, p. 22, HStAS.

8. Generalinspektor für Wasser und Energie to the Schluchseewerk, September 30, 1941, HStAS.

9. Badische Landesnaturschutzstelle, Der Geschäftsführer und Landesbeauftragte für Heimatschutz, March 20, 1942, p. 5, HStAS.

10. Schluchseewerk to the Generalinspektor für Wasser und Energie, September 9, 1942, Alwin Seifert to the Generalinspektor für Wasser und Energie, September 7, 1942, and Gutachten der Landesnaturschutzstelle Baden, November 30, 1942, p. 12, HStAS.

11. Memorandum of March 1943, HStAS.

12. Der Reichsforstmeister als Oberste Naturschutzbehörde to the Generalinspektor für Wasser und Energie, March 9, 1943, HStAS.

13. Schluchseewerk to the Ministerium des Kultus und Unterrichts als Höhere Naturschutzbehörde, October 27, 1942, p. 6, HStAS.

14. Der Reichsforstmeister als Oberste Naturschutzbehörde to the Generalinspektor für Wasser und Energie, March 9, 1943, HStAS.

15. See Thomas Zeller, *Straße, Bahn, Panorama. Verkehrswege und Landschaftsveränderung in Deutschland von 1930 bis 1990* (Frankfurt and New York: Campus, 2002); Helmut Maier, "'Nationalwirtschaftlicher Musterknabe' ohne Fortune. Entwicklungen der Elektrizitätspolitik und des RWE im 'Dritten Reich,'" in *Elektrizitätswirtschaft zwischen Umwelt, Technik und Politik. Aspekte aus 100 Jahren RWE-Geschichte 1898–1998*, ed. Helmut Maier (Freiberg: Technische Universität Bergakademie Freiberg, 1999), 129–66.

Total War?

16. Bernhard R. Kroener, Rolf-Dieter Müller, Hans Umbreit, eds., *Organization and Mobilization of the German Sphere of Power*, vol. 5 (pts. 1 and 2) of *Germany and the Second World War* (Oxford: Clarendon Press, 2000), 1:1012.

17. Der Reichsstatthalter in Baden to Generalforstmeister Alpers, December 23, 1942, p. 2, HStAS.

18. Willi Oberkrome, "Kerntruppen," in "'Kampfzeiten.' Entwicklungstendenzen des deutschen Naturschutzes im Ersten und Zweiten Weltkrieg," *Archiv für Sozialgeschichte* 43 (2003): 225–40.

19. Gutachten der Landesnaturschutzstelle Baden, November 30, 1942, HStAS.

20. Resolution of Badischer Schwarzwaldverein and other associations, January 1927, HStAS.

21. See Klaus Fehn, "'Lebensgemeinschaft von Volk und Raum.' Zur nationalsozialistischen Raum- und Landschaftsplanung in den eroberten Ostgebieten," in *Naturschutz und Nationalsozialismus*, ed. Joachim Radkau and Frank Uekötter (Frankfurt and New York: Campus, 2003), 207–24; and Gert Gröning and Joachim Wolschke-Bulmahn, *Die Liebe zur Landschaft. Teil III: Der Drang nach Osten. Zur Entwicklung der Landespflege im Nationalsozialismus und während des Zweiten Weltkrieges in den 'eingegliederten Ostgebieten'* (Munich: Minerva-Publikation, 1987).

22. Most recently, Hans-Werner Frohn used the emphasis on ideology to distract attention from the far more insidious complicity of conservationists in everyday practice; see his essay "Naturschutz macht Staat—Staat macht Naturschutz. Von der Staatlichen Stelle für Naturdenkmalpflege in Preußen bis zum Bundesamt für Naturschutz 1906 bis 2006—eine Institutionengeschichte," in *Naturschutz und Staat. Staatlicher Naturschutz in Deutschland 1906-2006*, ed. Hans-Werner Frohn and Friedemann Schmoll (Bonn-Bad Godesberg: Bundesamt für Naturschutz, 2006), 157–93.

23. See my argument for an institutionalist approach in "Confronting the Pitfalls of Current Environmental History: An Argument for an Organisational Approach," *Environment and History* 4 (1998): 31–52.

24. For more on this argument, see my *The Green and the Brown: A History of Conservation in Nazi Germany* (Cambridge: Cambridge University Press, 2006).

25. See Hans-Peter Ullmann, "Kriegswirtschaft," in *Enzyklopädie Erster Weltkrieg*, ed. Gerhard Hirschfeld, Gerd Krumeich, and Irina Renz (Paderborn: Schöningh, 2003), 220–32; Gerald D. Feldman, *Army, Industry, and Labor in Germany, 1914–1918* (Princeton, NJ: Princeton University Press, 1966); Jürgen Kocka, *Facing Total War: German Society, 1914–1918* (Cambridge, MA: Harvard University Press, 1984); Richard Bessel, *Germany after the First World War* (Oxford: Clarendon Press, 1993).

26. Deposit 1–2, file 506–382, fol. 28, Stadtarchiv Erfurt.

27. Ibid., fol. 42, Stadtarchiv Erfurt.

28. Depot B, C XVIII 3, Bd. 2, Nr. 6, pp. 44–45, Stadtarchiv Stuttgart.

29. *Verwaltungsbericht der Fabrik- und Handelsstadt Chemnitz 1914–1921* (Chemnitz, 1922), 267; *Haustechnische Rundschau* 19 (1914/15): 221.

30. Bundesrat, Tagung 1918, Drucksache Nr. 121, MWi 654, Hauptstaatsarchiv München (hereafter HStAM).

31. Rep. 30 Bln C No. 1878, p. 182, Brandenburgisches Landeshauptarchiv Potsdam; Bürgermeister Lucherberg to the Landrat of Düren, February 16, 1915, Regierung Aachen No. 14244, Hauptstaatsarchiv Düsseldorf (hereafter HStAD).

32. Verfahren 15, 4,14/1 V.A.4.bh, Staatsarchiv Bremen.

33. Verfahren 35, 4,14/1 V.A.4.bc, Staatsarchiv Bremen.

34. Verfahren 26, 4,14/1 V.A.4.bi, Staatsarchiv Bremen.

35. Rubner, "Das durch Artilleriegeschosse verursachte Fichtensterben," *Mitteilungen der Bayerischen Botanischen Gesellschaft zur Erforschung der heimischen Flora* 3, no. 13 (1916): 273–76.

36. 1–2/506–382, fol. 84, Stadtarchiv Erfurt.

37. Mülheimer Bergwerks-Verein to the Oberbürgermeister Essen, June 13, 1916, Bauaufsichtsamt No. 11444, Stadtarchiv Essen.

38. Cf. memoranda of August 20, 1915, September 2, 1915, November 6, 1915, and January 31, 1916, 611/4179, Stadtarchiv Duisburg.

39. Oberbürgermeister Duisburg to Firma Friedrich Rasche, August 9, 1915, and response of August 16, 1915, 306/584, Stadtarchiv Duisburg; Oberbürgermeister Duisburg to the Stellvertretendes General-Kommando VII. A. K. in Münster, November 14, 1916, Regierung Düsseldorf No. 34217, HStAD.

40. Konrad Wilhelm Jurisch, "Rauchfragen," *Chemiker-Zeitung* 40 (1916): 25–26, esp. 25.

41. Ilja Mieck, "Luftverunreinigung und Immissionsschutz in Frankreich und Preußen zur Zeit der frühen Industrialisierung," *Technikgeschichte* 48 (1981): 239–51; Ilja Mieck, "Die Anfänge der Umweltschutzgesetzgebung in Frankreich," *Francia* 9 (1981): 331–67; also see the overview in Michael Stolberg, *Ein Recht auf saubere Luft? Umweltkonflikte am Beginn des Industriezeitalters* (Erlangen: H. Fischer, 1994), 112–46.

42. See Frank Uekötter, "Das organisierte Versagen: Die deutsche Gewerbeaufsicht und die Luftverschmutzung vor dem ökologischen Zeitalter," *Archiv für Sozialgeschichte* 43 (2003): 127–50; Frank Uekötter, *The Age of Smoke: Environmental Policy in Germany and the United States, 1880–1970* (Pittsburgh: University of Pittsburgh Press, 2009).

43. Jahresbericht der Gewerbeinspektion für Bremen für die Jahre 1914–18, 3-G.4.g No. 32, p. 132, Staatsarchiv Bremen.

44. Kurt von Rohrscheidt, *Die Gewerbeordnung für das Deutsche Reich mit sämmtlichen Ausführungsbestimmungen für das Reich und für Preußen* (Leipzig: C. L. Hirschfeld, 1901), 65.

45. Ibid., 56.

46. Kriegsministerium to the Staatsministerium des Kgl. Hauses und des Äußern, May 30, 1918, MWi 654, HStAM.

47. Remarks of the State of Württemberg, June 10, 1918, HStAM.

48. Kriegsministerium to the stellvertretende Generalkommando, IV. Armeekorps, in Magdeburg, July 8, 1915, MH 14541, HStAM.

49. Bundesrat, Tagung 1918, Drucksache No. 121, MWi 654, HStAM.

50. *Reichsgesetzblatt 1918*, 1224.

51. The trading regulations also provided that companies would lose their license if they did not operate for more than three years. In the summer of 1917, the German government introduced legislation to make sure that firms that had closed due to the war would not lose their license. (Cf. Bundesrat, Session 1917, Drucksache No. 224, Abt. 233/26102, Badisches Generallandesarchiv Karlsruhe, and *Reichsgesetzblatt 1917*, 680.)

52. *Reichsgesetzblatt 1934*, part 1, 566.

53. Der Reichswirtschaftsminister und Preussische Minister für Wirtschaft und Arbeit to the Regierungspräsidenten, October 30, 1934, Regierung Aachen No. 12974, HStAD.

Total War?

54. Letter of the Reichswirtschaftsminister, February 20, 1941, Abt. 233/26103, Badisches Generallandesarchiv Karlsruhe.

55. See Uekötter, *Age of Smoke*.

56. *Journal of the Proceedings of the City Council of the City of Chicago for the Council Year 1917–18*, 1381–82. Cf. Chicago Department of Health, *Report for 1892*, p. 51; *Chicago Record-Herald*, November 8, 1901, p. 2, col. 4; November 9, 1901, p. 7, col. 5; February 8, 1902, p. 9, col. 1; and March 20, 1902, p. 9, col. 1.

57. Raul Hilberg, *Die Vernichtung der europäischen Juden. Band 2* (1961; rpt. Frankfurt: Fischer-Taschenbuch Verlag, 1990), 946.

58. Zentral-Bauleitung der Waffen-SS und Polizei Auschwitz to the Reichsanstalt für Wasser-, Boden- und Luftgüte, March 1, 1943, and response of September 23, 1943, R 154/48, Bundesarchiv (hereafter BArch).

59. See Hannah Arendt, *Eichmann in Jerusalem: A Report on the Banality of Evil* (1963; rpt. New York: Penguin Books, 1994).

60. Runderlaß des Reichswirtschaftsministers, R 154/12026, February 18, 1942, BArch.

61. Cf. R 154/11882, R 154/11896, R 154/11898, R 154/11903, R 154/11916, R 154/11951, R 154/11970, R 154/12015, R 154/12016, R 154/12017, R 154/12028, R 154/12042, R 154/12050, R 154/12051, R 154/12076, R 154/12091, R 154/12093, R 154/12100, R 154/12107, R 154/12121, R 154/12130, R 154/12136, R 154/12140, R 154/12146, R 154/12147, R 154/12149, R 154/12163, R 154/12171, all BArch.

62. Cf. proceedings in R 154/414, BArch.

63. Cf. R 154/11956, BArch.

64. Stadtgesundheitsamt No. 234, p. 57, Stadtarchiv Leipzig.

65. J. D. Küster Nachf. to the Oberbürgermeister, Abt. Baupolizei, Bielefeld, July 9, 1941, and responses of August 2, 1941 and August 21, 1941, MBV 038, Stadtarchiv Bielefeld. For more examples, see Frank Uekötter, "Polycentrism in Full Swing: Air Pollution Control in Nazi Germany," in *How Green Were the Nazis: Nature, Environment, and Nation in the Third Reich*, ed. Franz-Josef Brüggemeier, Mark Cioc, and Thomas Zeller (Athens: Ohio University Press, 2005), 101–28.

66. B 245/60, pp. 224–25; B 245/214, pp. 153–56, BArch.

67. See Karsten Runge, *Entwicklungstendenzen der Landschaftsplanung. Vom frühen Naturschutz bis zur ökologisch nachhaltigen Flächennutzung* (Berlin: Springer, 1998).

68. See Werner Weber and Walther Schoenichen, *Das Reichsnaturschutzgesetz vom 26. Juni 1935 und die Verordnung zur Durchführung des Reichsnaturschutzgesetzes vom 31. Oktober 1935 nebst ergänzenden Bestimmungen und ausführlichen Erläuterungen* (Berlin: H. Bermühler, 1936), 97–99.

69. Cf. proceedings in B 245/55, BArch.

70. *Reichsministerialblatt der Forstverwaltung* 7 (1943): 151.

71. Michael Wettengel, "Staat und Naturschutz 1906–1945. Zur Geschichte der Staatlichen Stelle für Naturdenkmalpflege in Preußen und der Reichsstelle für Naturschutz," *Historische Zeitschrift* 257 (1993): 355–99, esp. 396.

72. Reichsanstalt für Wasser- und Luftgüte to the director of Lurgi-Apparatebau, R 154/39, January 14, 1944, BArch.

73. Erich Naumann, *60 Jahre Institut für Wasser-, Boden- und Lufthygiene* (Stuttgart: Fischer, 1961), 62.

74. Reichsanstalt für Wasser- und Luftgüte to Wienersdorfer Asphaltfabrik Richard Felsinger, R 154/28, March 12, 1945, BArch.

75. Cf. R 154/85, BArch.

76. Cf. BR 1015/101, pp. 233–34, 253, HStAD; memorandum of January 7, 1943, Rep. 102 Abt. XIV No. 94, Stadtarchiv Essen; Hauptverwaltungsamt Kap. 28 No. 399, p. 7R, Stadtarchiv Leipzig; Vergleich zwischen der Forstgenossenschaft Lüttgenberg und den Unterharzer Berg- und Hüttenwerken GmbH, November 22, 1943, 12 Neu 15 No. 3869, Niedersächsisches Staatsarchiv Wolfenbüttel.

77. Regierung Arnsberg No. 1568, p. 414, Staatsarchiv Münster.

78. Reichsanstalt für Wasser- und Luftgüte to the Oberbergamt Breslau, R 154/11896, December 18, 1943, BArch.

79. Memorandum on the meeting of May 19, 1944, Regierung Arnsberg 6 No. 217, Staatsarchiv Münster.

80. B 245/214, p. 147, BArch.

81. Thomas Zeller, "'Ganz Deutschland sein Garten.' Alwin Seifert und die Landschaft des Nationalsozialismus," in Radkau and Uekötter, eds., *Naturschutz und Nationalsozialismus*, 273–307; and Uekötter, *The Green and the Brown*, 153–61. Even bearing these rivalries in mind, however, the proposal evokes memories of Victor Klemperer's mockery about the Nazis' penchant for organization (see Victor Klemperer, *LTI. Notizbuch eines Philologen* [Berlin: Aufbau-Verlag, 1947], 132).

82. See n. 22, above.

83. Günter Mader, *Gartenkunst des 20. Jahrhunderts. Garten- und Landschaftsarchitektur in Deutschland* (Stuttgart: Deutsche Verlags-Anstalt, 1999), 74–75.

84. R 18/3754, memorandum of August 1943, p. 2, BArch.

85. R 154/12027, report of April 26, 1942, p. 44, BArch.

86. Ibid., p. 45.

87. Ibid., p. 41.

88. Ibid., p. 45. See Kenneth Anders and Frank Uekötter, "Viel Lärm ums stille Sterben: Die Debatte über das Waldsterben in Deutschland," in *Wird Kassandra heiser? Die Geschichte falschen Ökoalarme* ed. Frank Uekötter and Jens Hohensee (Stuttgart: Steiner, 2004), 112–38.

89. R 154/12027, report of April 26, 1942, p. 46, BArch.

90. Reichsanstalt für Wasser- und Luftgüte to the director of Lurgi-Apparatebau, R 154/39, January 14, 1944, BArch.

91. See, e.g., Werner Durth, *Deutsche Architekten: Biographische Verflechtungen 1900–1970* (Munich: Deutscher Taschenbuch Verlag, 1992).

92. Letter of the Reichswirtschaftsminister, June 25, 1941, MWi 655, HStAM.

93. Arnold Heller, "Soll man in Deutschland das Abgasproblem, insbesondere bei Feuerstätten, gesetzlich regeln?" *Gesundheits-Ingenieur* 75 (1954): 386–92, esp. 390; Helmut Köhler, "Die durch die Technische Anleitung zur Reinhaltung der Luft zu erwartende rechtliche Situation," *IWL-Forum* 2 (1964): 307–12, 308; Entwurf, Technische Anweisung für die Genehmigung gewerblicher Anlagen (§ 16 GO.), B 149/10407, BArch.

94. See Charles E. Closmann, "Legalizing a *Volksgemeinschaft*: Nazi Germany's Reich Nature Protection Law of 1935," in Brüggemeier, Cioc, and Zeller, eds., *How Green Were the Nazis?*, 18–42, esp. 28.

95. Wettengel, "Staat," 383.

96. Uekötter, *Age of Smoke*.

97. Cf. Reichsminister des Innern to the Reichsanstalt für Wasser- und Lufthygiene, R 154/12026, March 5, 1943, BArch; *Haustechnische Rundschau* 20 (1915/16): 41.

98. Franz Neumann, *Behemoth: The Structure and Practice of National Socialism, 1933–1944* (1942; rpt. New York: Harper and Row, 1966), 78.

99. Landesanstalt für Wasser-, Boden- und Lufthygiene to the Reichswirtschaftsminister, R 154/12026, January 11, 1942, BArch.

100. Der Reichswirtschaftsminister to the Reichsstatthalter in den Reichsgauen Danzig-Westpreussen und Wartheland, die Landesregierungen, the Preußische Regierungspräsidenten and the Polizeipräsident in Berlin, June 25, 1941, MWi 655, HStAM.

101. Bernhard R. Kroener, "The Manpower Resources of the Third Reich in the Area of Conflict between Wehrmacht, Bureaucracy, and War Economy, 1939–1942," in Kroener et al., eds., *Organization*, 2:883–1070, esp. 807–8.

102. Cf. Bernhard R. Kroener, "Management of Human Resources, Deployment of the Population, and Manning the Armed Forces in the Second Half of the War (1942–1944)," in Kroener et al., eds., *Organization*, 2:775–1001.

103. Norbert Frei, *Der Führerstaat. Nationalsozialistische Herrschaft 1933 bis 1945* (1987; rpt. Munich: Deutscher Taschenbuch Verlag, 2002), 208.

104. *Reichsministerialblatt der Forstverwaltung* 7 (1943): 151.

105. Rede des Reichgesundheitsführers, Staatssekretär Dr. Conti, bei der Erweiterung der bisherigen preußischen Landesanstalt für Wasser-, Boden- und Lufthygiene zur "Reichsanstalt für Wasser- und Luftgüte," R 154/414, April 8, 1942, p. 10, BArch.

106. Zeller, *Straße, Bahn, und Panorama*, 204.

107. Hans Klose, *Fünfzig Jahre Staatlicher Naturschutz* (Giessen: Brühl, 1957), 35. I discuss this boom extensively in *The Green and the Brown*, 167–83.

108. On this phenomenon, see Frank Uekötter, "Sieger der Geschichte? Überlegungen zum merkwürdigen Verhältnis des Naturschutzes zu seinem eigenen Erfolg," *Schriftenreihe des Deutschen Rates für Landespflege* 75 (2003): 34–38.

109. See Douglas R. Weiner, *A Little Corner of Freedom: Russian Nature Protection from Stalin to Gorbachëv* (Berkeley: University of California Press, 1999); Douglas R. Weiner, *Models of Nature: Ecology, Conservation, and Cultural Revolution in Soviet Russia* (Bloomington: Indiana University Press, 1988); Judith Shapiro, *Mao's War against Nature: Politics and the Environment in Revolutionary China* (Cambridge: Cambridge University Press, 2001).

110. B 245/6, pp. 38–40, BArch.

111. Arbeitsgemeinschaft Heimatschutz Schwarzwald to Kultusminister Simpfendörfer, February 8, 1958, HStAS.

CHAPTER 6

WORLD WAR II AND THE AXIS OF DISEASE

Battling Malaria in Twentieth-Century Italy

Marcus Hall

In the summer of 1943 Allied troops began their invasion of southern Italy, and by April of 1945, the countryside from Sicily to the Roman Campagna was pockmarked by craters from bomb shells and artillery. This scarred landscape, along with villages and cities devastated by battle, became an ideal habitat for rodents and insects and perfect breeding grounds for mosquitoes. Soldiers and civilians across the peninsula were contracting typhus, tuberculosis, dysentery, and especially malaria, with many of these ills being spread by the vermin and parasites who were the immediate victors of every battle. As the dust settled and the front moved northward, villages such as Castel Volturno (which lies north of Naples) and then the Tiber Delta became the first places in Europe where health officials would experiment with dichloro-diphenyl-trichloroethane, or DDT, as a means to kill mosquitoes and thus control the spread of malaria (map 6.1). The next spring, following heavy winter rains, spray crews and airplane dusters began fumigating the soggy, wartorn countryside, testing out the miracle insecticide that had already proved itself so effective in the South Pacific. Warfare disrupted the ecosystems that kept malaria in check, but it also led to the development of new technologies that might enable eradication of this disease. In this Italian case of war and environment, I offer examples of how malaria modified combat and its outcomes, as well as examples of how warfare modified malaria. As we ponder other wars—past, present, and future—it behooves us to consider the multiple, though seldom-studied effects of linking belligerent humans with infectious microbes.

Malaria in Twentieth-Century Italy

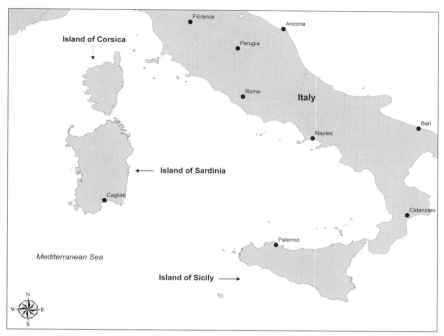

6.1 *Map of Sardinia, created by David Wilson, Center for Instruction and Research Technology, University of North Florida.*

Causes and Effects of Malaria

Soldiers have always shared their battlefields with disease. The chaos and unsanitary conditions that characterize every combat zone, combined with soldiers' lowered resistance to pathogens of all kinds, have meant that invading armies often tallied more losses from bugs than from bullets. Napoleon learned these facts of war through his defeat at Waterloo, where typhus disabled twice as many men as enemy fire. During the U.S. Civil War, one source claims that 1.2 million soldiers contracted malaria, with eight thousand of them succumbing to the disease. Gen. Douglas MacArthur understood the potential threat that an outbreak of malaria posed to U.S. troops in World War II when, midway through America's Pacific campaign, he told Army medical experts, "this will be a long war if for every division I have facing the enemy I must count on a second division in the hospital with malaria and a third division convalescing from this debilitating disease!" Within a few months of their July 1943 landing in Sicily, more than twenty-one thousand American and British soldiers had been hospitalized with malaria, surpassing in numbers their seventeen thousand comrades who had been wounded in battle. If it was bad strategy for the Germans to invade Russia in winter, it was also bad strategy for the Allies to storm Italy's pestilential coasts in summer.[1]

Italian officials kept careful watch over their nation's health. Infectious diseases such as malaria were considered impediments to progress, and Mussolini's ambitious program of Bonifica Integrale (a kind of Italian New Deal) included massive land-drainage programs that were meant to improve agricultural capacity as well as public health. Malaria was endemic to most of the peninsula's coastal marshes, and the drainage of these wetlands was seen as a crucial step toward the achievement of Italy's productive potential. Auspiciously, statistics like those compiled in the early 1960s by malariologist Alberto Coluzzi suggested that Italy was indeed winning its battle with malaria. According to his figures, which cover the sixty-three-year period from 1887 to 1950, the main deviations from the steady downward trend in malaria morbidity were the years of the two world wars. His graph shows a strong correlation between military conflict and increased risk of death from malaria (figure 6.1).[2]

Before pointing to the probable mechanisms by which warfare promoted malaria, one should realize that malaria statistics require interpretation. Variously termed *mal-aria* (bad air), *le febbre* (the fevers), and *paludismo* (swamp disease), the disease manifested itself in numerous ways and was hard to diagnose positively without specialized tests and equipment. Lacking access to laboratories in which the malaria parasite, or plasmodium, could be identified in a patient's blood sample, most country doctors in the first half of the twentieth century simply noted symptoms and palpated the patient's spleen—and then assumed that more distended spleens indicated greater malarial infection. Variable symptoms of the disease, ranging from fever and lethargy to nausea and chills, also made it difficult to single out malaria as the sole cause of sickness or death. In fact, plenty of *carriers* of the malaria plasmodium suffered few or no ill effects—physical or physiological—so that measuring *incidence* of malaria was especially difficult. Many health officials nonetheless trumpeted Italy's malaria declines, drafting downward sloping curves of malaria incidence that paralleled those of malaria mortality—with both curves showing wartime increases in malaria. Yet incidence numbers, like mortality rates, suggested a degree of certainty unsupported by the data.

Alberto Missiroli, one of Italy's leading malariologists during the war years, speculated even more widely when linking malaria rates to societal unrest. Offering a *longue durée* view of the Roman Campagna that reached across the twenty centuries since Christ's birth, Missiroli produced a graph that represented acute malaria with a line of three peaks that were mirrored by another line showing three agricultural declines. In the text accompanying this graph, Missiroli suggested that severe malaria foretold periods of agricultural disruption and social struggle, with these cycles recurring three times over the past two millennia. In its 1946 annual report the Rockefeller Foundation reproduced Missiroli's graph to help justify its ongoing investment in Italy's health-care system. Since the early 1920s the foundation had helped to finance Rome's new health institute, the Istituto Superiore di Sanità, and it was continuing to provide expertise and monies that would culminate in its heavily

Malaria in Twentieth-Century Italy

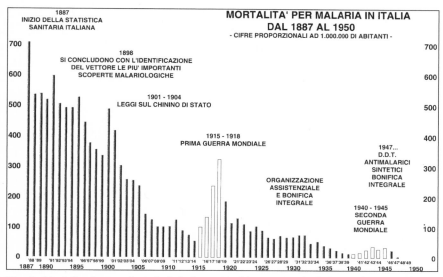

Fig. 6.1 Mortality from malaria in Italy from 1887 (the year in which national statistics began to be kept) to 1950, reproduced from Alberto Coluzzi, "L'eradicazione della malaria: Una sfida al mondo," Annali della Sanità Pubblica 22, no. 2 (1961): 241–53.

sponsored Sardinia Project, a massive postwar malaria-eradication effort. In reproducing Missiroli's graph, the Rockefeller Foundation was implying that the elimination of malaria would promote peace and accelerate economic development. But, on second view, this graph could just as easily be implying that promoting peace and development would rid malaria from the land. Was malaria a cause or an effect of social disruption, regional conflict, and warfare (figure 6.2)?[3]

An initial consideration of this question suggests that there is much more evidence to support the second proposal, namely that warfare exacerbated malaria. Not only did the water-filled bomb craters and tank tracks that zigzagged across Castel Volturno's countryside serve as mosquito breeding grounds, thus accelerating malaria transmission, but also a scattering of stagnant puddles could have large multiplier effects on mosquito populations. In malaria-control programs since the early 1900s, whether in temperate or tropical climes, even open water wells and the occasional pail or can of standing rainwater were scrupulously covered, or else sprayed with diesel oil to hinder the growth of mosquito larvae. Because a mosquito's maximum flying range is only one or two miles, the close proximity of battlefield puddles to human dwellings could play an important role in the spread of malaria. The adage that "malaria fled the plow" may have arisen through observing that malaria rates declined when farmers began draining swamps and leveling the land for growing their crops.

Yet just as war's environmental disruptions created opportunities for malaria

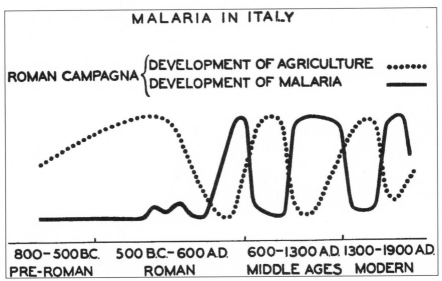

Fig. 6.2 The interrelationship between agricultural development and the incidence of malaria in Italy from pre-Roman to modern times, reproduced from Alberto Missiroli, "La Malaria nel 1944 e misure profilattiche previste per il 1945," Rendiconti dell'Istituto Superiore di Sanità (1944): 639.

to spread, war's social, political, and infrastructural upheavals also helped to transmit the disease. Critically, sustained battle often resulted in the disintegration of local health-care systems. For example, if hospitals and malaria clinics were not reduced to rubble, they usually lay far from the neediest patients. Antimalarial measures, such as the distribution of free or inexpensive quinine and atabrine, were also hindered by a state of war. These drugs offered a degree of prophylaxis (by protecting people from acquiring the malaria parasite) and of cure (by killing the parasite if contracted). Since 1900, the Italian government had subsidized large-scale, preventive quinine dosing in programs that were often referred to as *bonifica umana,* or human improvement. These quasi-eugenic measures of disease control were considered by their promoters to be one of the reasons why Italy had begun rolling back malaria, except during periods of war.[4]

Warfare also exacerbated malaria in more indirect ways. For example, when soldiers and civilians were forced to sleep in the open air because their dwellings and other forms of shelter were destroyed, they became more susceptible to disease-carrying mosquitoes. War refugees, too, flooding into Italy from Africa and Eastern Europe, many of them carrying the disease, served as reservoirs of the malaria plasmodium that hungry mosquitoes could distribute to the healthy population. World War II also disrupted the free trade of pyrethrum, one of the major insecticides employed for killing mosquitoes; the Japanese-dominated manufacture of pyrethrum from chrysanthemum flowers meant that the Allies found themselves scrambling to identify alternative insecticides.[5]

Some two decades after the Allied invasion of Italy, American malariologist Paul Russell reflected on the various ways by which war encouraged the spread of malaria, noting especially the habitat changes resulting from a war-ravaged landscape, the consequences of a disrupted health-care system, and urban refugees' heightened exposure to mosquitoes in the countryside. Russell, a Rockefeller Foundation health officer who directed the Allied Commission's malaria control efforts, also witnessed the aftermath of sabotaged public works along the Roman coast, where the retreating Nazis flooded fields by destroying irrigation pumps and obstructing drainage canals. When Alberto Missiroli observed that these newly inundated areas were nurturing a local malaria epidemic by the summer of 1944, he judged these coastal landscapes to have regressed to their nineteenth-century conditions. Missiroli warned that war's activities had returned these marshlands to their preindustrial, miasmal state.[6]

A closer reading of the wartime records, however, indicates that the Nazis' destruction of the Tiber delta's pumps is better understood not as sabotage but as biological warfare. The retreating Nazi troops realized that inundating this delta would multiply their enemies' risk of contracting malaria. A half meter of standing water on Rome's coastal plains not only hindered travel by foot and vehicle; it recreated ideal biological conditions for breeding *Anopheles labranchiae*, the local mosquito vector responsible for transmitting the malaria plasmodium.

According to Missiroli's official communications, the German command gave orders on October 9, 1943, to turn off all drainage pumps in the reclaimed Maccarese area of the northern Tiber delta. Over the next two weeks, most other drainage pumps across the remainder of the delta were also stopped, and in some cases, the pumps were actually reversed to begin inundating previously drained areas. Drainage canals were also blocked to promote greater flooding, while key levees were breached to allow saltwater to flow into reclaimed areas; entomologists knew that moderately salty water favored the development of mosquito larvae. In fact, German malaria experts Erich Martini from the University of Hamburg and Ernst Rodenwaldt from the University of Heidelberg had been sent to the delta that autumn to oversee inundation operations. Missiroli reported that by the first of December, 3,000 hectares in the Tiber delta had been submerged, together with another 6,000 hectares in the nearby Agro Pontino to the south. Nazi officials were perfectly cognizant of the malaria problems that their enemy had confronted a few months earlier in Sicily, and with the Allies now pushing at the gates of Rome, the Nazis were hoping that malaria-carrying mosquitoes could again be recruited to their side. No wonder Paul Russell often referred to malaria as the "Plasmodium-arthropod Axis."[7]

For Missiroli the flooding of the Tiber delta represented more than a way to fend off invading armies. Surviving correspondence and a diary suggest that

Missiroli had himself collaborated with the German authorities, not only helping them to properly flood the area but also fostering a malaria epidemic that he realized would be useful for carrying out future investigations of the disease; more patients would allow for more remedies to be tested. Always the scientist, Missiroli saw a newly flooded delta as an excellent laboratory for advancing the field of malariology.

It should be pointed out that Missiroli enjoyed a long working relationship with his German colleague Martini, with whom he had jointly authored scientific papers. Although it is unclear whether Missiroli or Martini was the first to call attention to the investigative advantages of flooding the Tiber delta, by late August of 1943, Missiroli was already suggesting to the Maccarese Reclamation Company that his own Laboratory of Parasitology be put in charge of the area's malaria-control efforts. Six weeks later the Maccarese area's main drainage pumps were turned off, and shortly thereafter Missiroli fired all but two of the personnel responsible for carrying out mosquito-control operations. In mid-November, Missiroli toured the flooded regions with Martini; together they recommended that the pumps remain turned off for "scopi bellici"—military reasons—while cautioning Italian health authorities to prepare for the coming malaria season. Another of Missiroli's colleagues, Alberto Coluzzi, would note in his diary that Missiroli helped to mastermind the Maccarese flooding.[8]

That next summer, after the June arrival of Allied troops in Rome and nearly a year after the Italian government's official surrender on September 8, 1943, Missiroli was blaming the Germans for the flooding and the resulting malaria epidemic. While he proposed that the Allies begin fitting the delta's houses with mosquito screening, he was much more interested in trying out the newly discovered insecticide dichloro-diphenyl-trichloroethane. In fact, so intent was he on testing out DDT that he decided to temporarily suspend all distribution of antimalarial drugs in the Maccarese. He reasoned that medications such as atabrine could quickly resolve that season's rising malaria problem, but to distribute them would make it difficult to interpret the efficacy of DDT spraying. Missiroli rationalized his experimental priorities by explaining that most of the new malaria cases arising in the Maccarese would be of the relatively innocuous third-order type—at least initially—and so could be ignored when designing DDT studies. Unfortunately for many residents of the delta, Missiroli's experiments also required that several other areas remain unsprayed and that their inhabitants remain untreated to serve as controls. In these unsprayed areas and untreated control populations malaria would temporarily "assume vast proportions" over the next twelve months.[9]

It should also be pointed out that Missiroli came from a long tradition of experimental malariology that sacrificed the needs of the few for the potential benefit of the many. Missiroli's laboratory had carried out other investigations in malaria-infested regions of Calabria and Sardinia, whereby recognized malaria remedies were withheld for testing a new procedure or treatment. For example, in the 1920s the countryside surrounding the Sardinian villages of

Posada and Portotorres became the first sites in Europe where Paris Green was used to control mosquito larvae; while this arsenic powder demonstrated deadly efficacy against the malaria vector, it also showed plenty of dangerous side effects on animals and humans. Again several years later, inhabitants of Posada became the first Italians to be administered a drug called Plasmochine instead of quinine as a prophylactic against the malaria parasite. Not surprisingly, when George MacDonald, director of Britain's Ross Institute, contacted Missiroli in 1946 about testing out yet another antimalarial called Paludrine, Missiroli once more suggested that the experiment be conducted in Posada. "The trials should take place in Sardinia," Missiroli wrote back, "since fortunately malaria has practically disappeared from central Italy."[10]

Entranced by the euphoria of medical progress, many malaria investigators downplayed or ignored ethical considerations in their experiments. It is appalling but not altogether surprising that prisoners and mental patients were routinely inoculated with malaria parasites and were then administered experimental drugs or subjected to experimental procedures in order to test possible cures. Missiroli's laboratory worked closely with Rome's psychiatric clinics, where patients with advanced syphilis were institutionalized and then inoculated with malaria. By inducing high fevers, this malaria therapy, as it was called, helped to attenuate the psychotic effects of syphilis and provide some relief from that disease until these syphilitic patients were brought out of their malarial stupor two or three weeks later with the administration of quinine. Missiroli and his colleagues, along with several other American and European malariologists, considered malaria therapy an ideal method for testing the efficacy of novel antimalarial remedies and drugs. Although "Smalarina" and "M.3," for example, were just two substances that may have demonstrated certain advantages over other medications used at Rome's Asylum of Santa Maria della Pietá, most of Missiroli's trials undoubtedly showed that the patients undergoing malaria therapy would have been better off simply taking quinine.[11]

War accelerated medical testing by multiplying opportunities for experimentation and by lowering ethical standards. From the perspective of generals and colonels, the threat of epidemics spreading through the front lines demanded immediate action. Medical researchers scrambled for cures, sometimes abandoning their Hippocratic Oath. A case in point is the famous typhus threat at Naples in the winter of 1943–44, which saw the first widespread civilian use of DDT. Generally celebrated as an Allied triumph in which military doctors intervened at the eleventh hour to quell a major louse outbreak and so extinguish the incubating typhus epidemic, this episode of preventive medicine relied on massive civilian spraying of a barely tested pesticide. Each week, spray nozzles were pushed under the arms and into the crotches of hundreds of thousands of men and women, infants and elderly: "the sight of persons on the street with powdered hair and clothing was too common to cause comment." Some three million separate DDT dustings were performed on Neapolitans over a six-month period. The toxic aftermath could have been catastrophic.[12]

Nonetheless, Gen. Morrison Stayer and Col. William Stone, after consulting health experts such as the Rockefeller Foundation's Fred Soper, decided on the spot that the risk of epidemic typhus outweighed the potential dangers of systemic poisoning by DDT. And dangers there were. As historian Edmund Russell reveals, one researcher who had conducted trials with DDT a few months earlier at the U.S. government's Orlando laboratory noted that "The preliminary safety tests, made with full strength DDT, had been somewhat alarming. When eaten in relatively large amounts by guinea pigs, rabbits and other laboratory animals, it caused nervousness, convulsions or death, depending on the size of the dose."[13] Needless to say, it is doubtful that many of the Neapolitans who lined up for their biweekly delousing sessions were shown the results of the Orlando study.

A review of Italy's recent military history therefore makes it clear that war not only spread infectious diseases, it also promoted the search for cures, often under abbreviated safety protocols. Bomb craters and reversed drainage pumps aggravated the malaria threat while pesticide development and human experimentation favored its control. Or, as Paul Russell saw it, war conditions spread malaria through "troop mobility and dispersion, necessarily based on tactics and not on sanitary conditions; a great deal of nocturnal activity; difficult logistics, especially in combat zones; enemy action, mines, and booby traps; and combat tension when the chief concern is not malaria control but immediate life and death." He also believed that war conditions aided the control of malaria by providing "complete authority of the commanding officer, uniformity of living habits of the personnel, and ample anti-malaria funds and supplies." By broadening biological knowledge of mosquitoes and by accelerating the understanding of the plasmodium, Russell concluded, "there can be no doubt that antimalaria activities of World War II constituted a prime factor in the development of the present move for worldwide malaria eradication." In other words, the state of emergency that is war channeled scientific intelligence and energy, multiplying the benefits of human genius along with the penalties of human folly. The linked history of war and disease is a history of greater means producing greater ends, a history of greater human desperation countered by greater human hope.[14]

The Sardinian Project

By 1945, the Fascists and Nazis had surrendered, but typhus and malaria had not. Following the massive DDT campaign in Naples, the Rockefeller Foundation, with assistance from the Italian government and United Nations relief monies, planned an even bigger and more ambitious antiarthropod campaign on the island of Sardinia. This time the enemy was not the louse but the mosquito, and the resulting battle was even closer to a real military operation. From 1946 to 1951 a special corps—the Ente Regionale per la Lotta Anti-Anofelica di Sardegna, referred to by the acronym ERLAAS—was formed to exterminate malaria-carrying mosquitoes throughout the island, using jeeps, airplanes, spray guns, tons of DDT, and thousands of uniformed men. As in Naples, this project also targeted the vector rather than the parasite, but it was aimed at eradication rather than control. As Fred Soper explained: "Mass dusting in Naples was not a louse eradication project, it was a typhus control measure for reducing the louse population to a point where typhus transmission would cease." The goal of ERLAAS was to exterminate every last malaria-carrying mosquito on an island the size of Vermont. As a *lotta*—or struggle—ERLAAS's undertaking was a paramilitary operation. It made the Naples project look easy.[15]

Although Soper was the mastermind of the Sardinia Project, DDT was its chief weapon. Now armed with what Soper called "an almost perfect insecticide," the Rockefeller Foundation hoped to turn Sardinia into a demonstration site in the worldwide fight against malaria. If *Anopheles* could be eliminated locally, it might be eliminated globally. Although scattered mosquitoes were still buzzing in Sardinia at the end of the project, Sardinians and the Sardinian environment would never be the same again.[16]

There were in fact various precedents for the Sardinian Project. A few months before the dusting of DDT in Naples, Soper and members of the Pasteur Institute had supervised smaller louse-killing projects in Algiers, including those at various prisoner-of-war camps, Arab villages, and the Maison Carrée Prison. After observing the "striking results" obtained through these initiatives, the Allies invited the Rockefeller Foundation to establish a Malaria Control Demonstration Unit for testing the efficacy of DDT in killing mosquitoes at Castel Volturno, and a few months later, at the Tiber delta. By summer of 1945, Missiroli's laboratory was overseeing other fumigation projects in the battlefields of Cassino, where sprayers tried out various concentrations of DDT as well as coating the insides of houses and animal shelters. By war's end, Italian health officials were spraying DDT up and down both of the country's coasts and, in Sicily, using the wonder insecticide to kill flying mosquitoes, to kill mosquito larvae in water, and—as an added bonus—to kill bothersome houseflies. Only Sardinia was kept out of their jurisdiction; as Italy's most malarious place, Sardinia was reserved for the investigative trials of ERLAAS.[17]

Besides these DDT-spraying precedents, there were also vector-eradication precedents. In 1938–40 in Brazil, Soper had supervised the successful eradication of another malaria vector, *Anopheles gambiae*, by the strategic use of Paris Green. In 1942 in Egypt, he again directed the eradication of *gambiae* mosquitoes along the upper Nile using similar methods. Such projects really represented *local* eradication, as this species continued to thrive outside of spray areas. Soper's dream, though, was complete eradication, a term once reserved for disease but increasingly applied to insects. According to Soper, eradication was "the ultimate in species reduction and implies the world-wide extermination of a species." In the case of Sardinia, the elimination of a disease vector across an entire island was expected to demonstrate in microcosm the prospects of global eradication. During the same years as the ERLAAS initiative, plans were also made for eradicating mosquitoes on the islands of Cyprus and Crete. But the Cypriot and Cretan campaigns would not be as thorough or as deadly as the Sardinian one.[18]

In 1948 the London-based Shell Petroleum Company sent a film crew to Sardinia to take footage of the mosquito-eradication project for a planned documentary recording this historic malaria experiment. Proposed by Rockefeller Foundation leaders and financed by Shell, the thirty-five-minute film, titled *The Sardinian Project*, captured the warlike tenor of this eradication effort. Between scenes of men marching across the countryside shouldering spray tanks, fumigating wells and streams, digging canals for draining swamps, even exploding dynamite to hasten excavation or using flamethrowers to clear brush, the film's narrator explained that ERLAAS was a true military operation. "In November 1947, the first phase of the all-out campaign opened," said the narrator. When the five-year project ended, over $11 million had been spent and thousands of men had been employed to mix and spray, dig and drain, fix, plan, check, and supervise. The method was straightforward enough: locate all known water patches on the island; drain them or spray them with DDT (or Paris Green); spray DDT on the inside walls of every dwelling, human or animal; check all such places for surviving mosquito larvae or adults, and, if any were found, spray again. Official records indicate that more than a million separate water sources were located and sprayed. Cars, trucks, helicopters, airplanes, boats, maps, boots, and masks equipped an average seasonal crew of ten thousand workers—thirty-two thousand at its height—almost all Sardinians, except for project administrators, who were mostly Americans or other foreign experts employed by the Rockefeller Foundation. *The Sardinian Project* was followed by *Battaglie di Pace* (Battles of Peace), an Italian short that would be projected in cinemas across the island. Then, in 1950, *Adventure in Sardinia* was released, which was a lighter, more entertaining version of Shell's first film. Its narrator proclaimed that "ERLAAS was indeed an army."[19]

War metaphors permeated more than these documentary films. Throughout ERLAAS publications, spray-gun-toting soldiers were shown shooting at winged enemies. In a monthly report, *il Pericolo Sovrastante* (the Overwhelming Danger) was gunned down into the sea to become *il Nemico Fulminato* (the Annihilated Enemy). Such images bear out Edmund Russell's observation that creepy crawlers were often represented as humanoid enemies, particularly when real wars were close at hand and when one could kill by the press of a button. As in the United States, DDT in Sardinia became the insecticide bomb in the water well, the mushroom cloud under the sink. As Russell shows, analogies between insects and humans, between insecticide gas and nerve gas, were simply too close to be ignored by those searching for remedies to the day's immediate crises. Experts in pest control were sometimes recruited for human weapons research. A war-obsessed world saw its other struggles in warlike terms. Little surprise, then, that for ERLAAS administrators, enemies could be Nazi-Fascist as well as Plasmodium-arthropod (figure 6.3).[20]

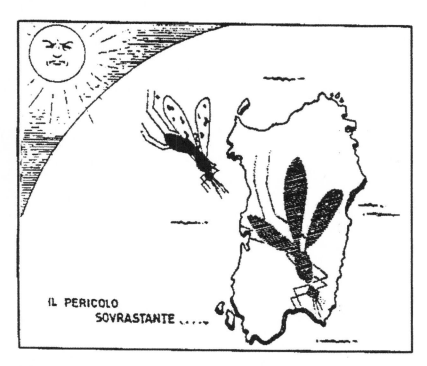

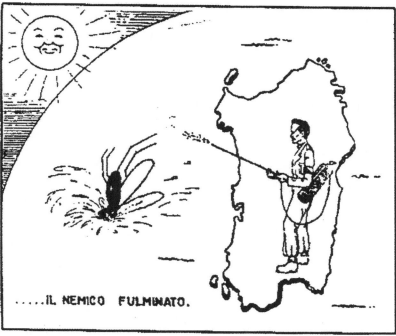

Fig. 6.3 Two-panel cartoon from a pamphlet distributed in Sardinia by ERLAAS (Ente Regionale per la Lotta Anti-Anofelica di Sardegna) c. 1948. The first panel dubs the mosquito a "pericolo sovrastante," or impending danger, and the second panel depicts it as the "nemico fulminato," or the enemy annihilated.

But just as all analogies risk overextension, it became increasingly apparent that mosquitoes did not pilot warplanes, nor did malaria wear a swastika. ERLAAS was ultimately an entomological experiment: an investigation for testing the feasibility of *Anopheles* eradication. ERLAAS was not an act of war, nor was it a public-health measure for controlling malaria. At the beginning of the project, eradicating the mosquito and controlling the disease seemed to require roughly the same approach: a generous spraying of DDT. Yet as the project wore on, ERLAAS supporters came to realize that these two goals could require dramatically different procedures, especially as mosquitoes kept buzzing while malaria cases kept dropping. Marston Bates, a Rockefeller Foundation ecologist who kept careful watch on the Sardinian Project, later suggested that ERLAAS participants "were sometimes uncertain as to whether they were conducting an experiment or implementing a public-health measure." Sardinian Project supporters pointed out that investing in Sardinia's long-term health also injected much-needed cash into this island's desperately poor economy. But there were surely cheaper ways to control malaria, and there were clearly better ways to spend the money: on hospitals and schools, or on railroads and harbors, for example. While the coming of ERLAAS meant that shepherds-turned-DDT-sprayers could finally "carry coins in their pockets," as one sprayer put it, it is also true that such workers might well have funneled their efforts into more crucial work.[21]

In fact, those in the know understood that malaria could be extinguished with only moderate DDT spraying. In 1944 trials at the Tiber delta, both Missiroli and Soper had observed spectacular malaria retreats with even cursory DDT coverage. Missiroli predicted that malaria could be wiped out across all of Italy in just two to three years, and in fact Italy's health ministry would do just that upon adopting Missiroli's recommendations—except in off-limits Sardinia. Even though Sardinia's rates of malaria incidence and malaria mortality declined just as rapidly as those on the mainland, ERLAAS sprayers pushed on for several more years, refilling and spraying anew—searching out every last *Anopheles labranchiae* even as mosquito DDT resistance set in. In only the second year of the Sardinian Project, ERLAAS director John Logan made the job of malaria control sound trifling as compared to vector eradication. When he answered queries from French health authorities who were confronting the rising malaria problem on the nearby island of Corsica, he calculated that just one-third to one-fourth as much DDT would be needed if their goal was *merely* to control malaria instead of to eradicate mosquitoes.[22]

True, there was the possibility that the strategy of control would require continual respraying over the years to keep malaria in check, so that vector eradication would represent the best long-term solution to malaria. But experience across the rest of Italy, in Greece, and in the South Pacific was already suggesting that light DDT spraying was sufficient to break the malaria cycle. An 80 percent reduction in the vector population was enough to quell a malaria epidemic. In fact in the American experience with malaria, as Margaret Humphreys argues, zealous DDT spraying in the southern states during the 1940s was akin to "kicking a dying dog," so imminent was the demise of malaria in the United States through other control methods. It seems that Sardinia's dying dog was also being kicked—and hard. Immediately before the war and before the advent of DDT, levels of malaria in Sardinia had dropped to an all-time low; and even with temporary wartime relapses, malaria was assuredly on the way out on this island, with or without Paul Müller's insecticide discovery.[23]

Sardinians then and Sardinians now hardly realize that ERLAAS sprayers were little more than underpaid lab assistants. "Today, thanks to DDT," announced a recent commemorative exhibition at the elementary school of Birori, Sardinia, "malaria has disappeared from Sardinia and from other temperate regions." Ex-ERLAAS-sprayer Giuseppe Foeddu, when interviewed fifty years after the Sardinian Project, expressed gratitude to the Americans: "thanks to them, malaria had been snuffed out." Elderly Giuseppe Flore, another ex-sprayer in the project, offered a perspective shaped by fifty years of hindsight when he remarked that "the word DDT is synonymous with savior from disease, with well-being, and for me a steady job with an income.... Luckily they invented it, otherwise the feared and terrible malaria would still have been with us, causing many deaths, especially of children and other defenseless and vulnerable individuals." But Fascist-era quinine programs and marsh-drainage schemes, together with new mosquito-proof housing were already driving malaria away when the men with spray tanks appeared on the horizon. Most Sardinians continue to argue that a little DDT in their grandparent's (and their own) blood is better than suffering from malaria. Or as elderly Mariantonia Loddo from Ortueri expressed her views: "Yes, I remember the anti-malaria campaign. They entered people's houses telling us that it wasn't poisonous. And now see what they say! Well... they did it anyway because in those days it was a true and real epidemic." She and most others who witnessed the project rarely suggest that DDT might have been avoided altogether, or even that the spraying might have been limited to as little as one-quarter of the amount of insecticide actually applied.[24]

Indeed, health experts were in wide disagreement about how malaria should be eliminated in Sardinia. The first director of ERLAAS, John Kerr, resigned just a few months after the start of the project, declaring that island-wide mosquito eradication was an unrealistic goal. As Kerr exclaimed to Fred Soper, "In my opinion, the organization of comprehensive anti-larva work in the portion of Sardinia which has an elevation of up to 1,000 meters is an impossible task. Call me a pessimist if you will, but the word *impossible* is in my vocabulary, and I intend to keep it there."[25]

Although plenty of other malariologists, including Italy's Missiroli, believed that the Sardinia Project could and should be carried out to completion, other Italians lobbied for vector control over vector eradication. Several of Sardinia's own experts, including two of the island's four Medici Provinciali, called for a return to "Italian methods" in lieu of the new "American methods."

Italy's traditional programs of swamp draining and oiling, along with widespread quinine distribution, became more appealing as reports accumulated about DDT-poisoned bees, fish, livestock, and even people. ERLAAS opened a public-relations office and found itself a defendant in numerous lawsuits claiming property damages. That the courts rapidly and easily dismissed most such claims indicates that ERLAAS was not reluctant to exert its influence with friends in high places.[26]

Historian Eugenia Tognotti argues that the Sardinian Project was part and parcel of the political schism dividing the island during the postwar years. The Christian Democrats were vying with the Communists as the majority party in this war-torn land, with ERLAAS sometimes being portrayed in newspapers as pro-American and anti-Communist. Tognotti's most compelling evidence is a letter from Britain's Lord Boyd Orr, director of the United Nations' Food and Agriculture Organization (FAO) and a 1949 Nobel Prize winner, addressed to America's ambassador to Britain about the desirability of maintaining the Rockefeller Foundation's presence in Sardinia after the end of the mosquito project. Orr recommended that the Rockefeller Foundation stay on to develop an islandwide economic plan in order to quell the rising pro-Communist sentiment there. But while there is little question that ERLAAS found itself in the midst of party and regional struggles (in which Sardinia emerged as a semiautonomous region), there is scant evidence to suggest that ERLAAS was designed by the United States as a strategic move in an escalating Cold War. America's military presence continued to grow in Sardinia, but this presence arose as a response to subsequent events unfolding in the rest of Europe rather than as a premeditated political maneuver thinly disguised as a malaria project. While Sardinia did serve as an Allied airbase for bombing northern Italy, it is unlikely that ERLAAS was, as some critics claim, really a front for establishing a kind of gigantic, permanent aircraft carrier for safeguarding American interests in the Mediterranean.[27]

Other interpreters of the Sardinian Project suggest that it should be considered primarily an entomological experiment, or a public-health project, or perhaps even a pork-barrel relief effort, rather than a tentacle of American foreign policy. Championed by an elite corps of technocrats acting with somewhat too much hubris and hegemony, the Sardinian Project was all of these things. There are many ways to explain the Rockefeller Foundation in Sardinia, most of which depend on the fortuitous encounter of pests and disease with politics and technology. As Marston Bates notes in his preface to the project's final report, "Each reader will probably draw his own moral from the tale; but that is the beauty of it. The facts are here, for the thoughtful reader to ponder in terms of his own interests, prejudices and developing plans."[28]

Yet together with these other interpretations, the Sardinian Project must also be viewed as an intersection of war and environment. Armed combat and the natural world must be considered in tandem when explaining the origins and effects of Sardinia's DDT dousing. The Allied troops who stormed Italy's malaria-infested beaches set the stage. A rich Italian tradition of experimental malariology, coupled with the discovery of a new pesticide, the Rockefeller Foundation's desire to spray it, and the United Nations' willingness to pay for it all produced the largest mosquito battle in history. Sardinia's ecosystems, its economy, its inhabitants' blood chemistry were forever altered. Malaria disappeared on the island, with warfare first worsening the epidemic and later providing the resources that led to its demise. After five years and 5.7 million liters of DDT solution, not just mosquitoes, but bees, fish, birds, and livestock were all poisoned. Meanwhile, Sardinia's massive DDT spraying also allowed some of the island's wetlands to remain wet, as humans no longer drained them in order to combat malaria. Copious DDT likewise meant that Sardinians were spared the nauseous side effects of quinine. War and its aftermath modified the land and its inhabitants, for bad as well as good.[29]

Today the Second World War is still etched across Sardinia's landscape. New buildings in Cagliari have been constructed on the rubble of houses destroyed during air raids. There are the old and the new military airbases where NATO's supersonic aircraft refuel and plan missions. There is the hundred-square-mile military proving ground in the southwest peninsula of Teulada where NATO war games leave tank tracks and bomb craters in the rolling grassland alongside beaches gouged by landing crafts. But while mosquitoes once again breed copiously in the puddles formed by these military maneuvers, such mosquitoes no longer carry the malaria plasmodium. The Sardinia Project's three-to-four-order overkill of malaria has left a rebounding population of *Anopheles* with an appetite for blood meals but scant chance that they will again threaten the islanders with the disease. Bomb craters as well as plasmodium-free mosquitoes are part of Sardinia's World War II legacy.

DDT Legacies

The simple story of war and malaria in modern Italy centers on how military emergencies accelerated the development of DDT, which led to quick eradication of the disease. Fuller studies of this case of war and malaria will explore how grain harvests plummeted when the enemy flooded fields for incubating new malaria epidemics, how novel pesticides modified ecosystems as well as human health, how military doctors could begin paying more attention to shrapnel wounds than to malarial fevers, and how insect battlefields helped to pave the way for NATO's modern infrastructure. War and malaria—these subsets of culture and nature—must be understood for what each did to the other, and for what the resulting changes would mean for each. At one point in his classic history of Italy, Dennis Mack Smith declares that malaria eradication may be "the most important single fact in the whole of modern Italian history." Yet one must also consider that Italy's victory over malaria was intimately linked to warfare, and that the methods of battling malaria were distinctly warlike.[30]

Marston Bates, the well-known ecologist and observer of the Sardinian Project, went on to consider larger questions about humanity's place in the natural world. Together with Carl Sauer and Lewis Mumford, Bates would help to organize the famous 1955 Marsh Festival that convened seventy of the day's leading environmental scholars to discuss "Man's Role in Changing the Face of the Earth." In his own commentary on this meeting, Bates called special attention to the role played by war in producing earthly changes. "Certainly, war has been a tremendously important agency in this process," he declared. Yet the subject of war, he pointed out, had attracted almost no discussion at the meeting. "Even though we have talked about war so little, clearly it has been hanging over our minds all through our discussions, as it hangs over the minds of all men in the Western world these days." Although Bates offered few illustrations of war's dramatic environmental effects, he implored his colleagues to begin tracing and explaining such effects. Surely Bates's own reflections on Europe's war-torn condition and on military spin-offs such as the Sardinia Project convinced him of how drastically warring humans might transform the ecosystems on which they depend.

Indeed, Bates's comments seem especially pertinent in regard to his own specialty of entomology. For example, one insect survey in Sardinia reveals that only sixteen out of twenty-four black fly species (or *Simulli*) survived the island's massive DDT episode. Although but a single example in one small corner of the world, this insect survey reflects the degree to which war's chemicals and war's relief programs can alter the fabric of nature.[31]

But war does not always destroy nature or ruin the land. It is true that war-developed pesticides seriously threatened Sardinia's fish-farm industry during the summers of 1947 and 1948, when mullet being raised in coastal marshes died by the thousands after airplanes blanketed these areas with DDT. Elsewhere in Sardinia, however, antimalaria squads transplanted gambusia fish into various streams and ponds in the hope that these North American imports would slurp up mosquito larvae and thus limit mosquito populations. Mullet stocks plummeted while gambusia numbers soared. While DDT spraying assuredly killed innumerable arthropods that relied on Sardinian marshes, the advent of powerful pesticides meant that engineers no longer drained marshes for health reasons. Innumerable aquatic creatures in these marshes owed their lives to DDT. It is more accurate to say that Sardinia's postwar malaria project remade rather than ruined the local ecosystems.[32]

The relationships between war and malaria, between humans battling humans and humans battling pathogens, are complicated, however. Today 23 percent of Sardinians are susceptible to favism, a hereditary enzyme deficiency involving red blood cells that, while making it difficult (or fatal) to digest legumes such as fava beans, also provides some protection against malaria. Evolutionary pressures over the centuries meant that highly malarial areas gave a selective advantage to people with favism. Now that Sardinians no longer confront malaria, there are no advantages for them to be born with favism; so this genetic condition is disappearing from the local gene pool. Sardinia's war on mosquitoes modified local ecosystems as well as human genes. Wars produce pesticides; pesticides control pathogens; pathogens reorder human DNA. Rockefeller Foundation experts reached into Sardinia to pull out the malaria threat and, like John Muir, found it hitched to everything else in the universe.

War comes in many forms, involving guns or words, physical injuries or psychological threats. War is armed and intense struggle, accelerating the interaction of the human and nonhuman, intentionally and unintentionally. War spreads disease and fashions remedies for its cure. In the case of Italy, people were beneficiaries as well as victims of such remedies. The tentacles of war turned Sardinia's flora and fauna, its mountains and coasts, into an ecological laboratory as well as a political proving ground, changing the face of the earth as well as the structure of the human genome.

Notes

1. See John R. Meyer, "Pests of Medical Importance," http://www.cals.ncsu.edu/course/ent425/text18/medical.html, accessed November 10, 2007; Fiammetta Rocco, *Quinine: Malaria and the Quest for a Cure That Changed the World* (New York: HarperCollins, 2003), 178–79; Paul Russell, Introduction to *Communicable Diseases—Malaria*, vol. 6, *Preventive Medicine in World War II,* comp. Office of the Surgeon General, Department of the Army, 9 vols. (Washington, DC: Government Printing Office, 1955–), 2, 9, 262.

2. See Alberto Coluzzi, "L'eradicazione della malaria. Una sfida al mondo," *Annali della Sanità Pubblica* 22, no. 2 (1961): 241–53.

3. Raymond Fosdick, *The Rockefeller Foundation: A Review for 1946* (New York: n.d.), 19; Missiroli's graph is reproduced in many of his publications, including Alberto Missiroli, "La Malaria nel 1944 e misure profilattiche previste per il 1945," *Rendiconti dell'Istituto Superiore di Sanità* (1944): 639.

4. Gilberto Corbellini and Lorenza Merzagora, *La Malaria: Tra Passato e Presente* (Rome: Museo di Storia della Medicina [Museum of the History of Medicine], 1998), 63.

5. Margaret Humphreys, *Malaria: Poverty, Race, and Public Health in the United States* (Baltimore: Johns Hopkins University Press, 2001), 147; William M. Tsutsui, "Landscapes in the Dark Valley: Toward an Environmental History of Wartime Japan," in *Natural Enemy, Natural Ally: Toward an Environmental History of War,* ed. Richard P. Tucker and Edmund Russell (Corvallis: Oregon State University Press, 2004), 209.

6. Paul Russell, Introduction, 5; Alberto Missiroli to the Direttore Generale dell'Istituto Superiore di Sanità [hereafter ISS], November 3, 1943, busta 6, fasc. 19, Laboratorio di Parassitologia, ISTISAN, Archivio Centrale dello Stato, Rome. See also Frank Snowden, *The Conquest of Malaria in Italy, 1900–1962* (New Haven, Conn.: Yale University Press, 2006).

7. Alberto Missiroli to the Direttore Generale dell'ISS, November 29, 1943, and Alberto Missiroli to the Sottosegretario di Stato, August 24, 1944, both in busta 6, fasc. 19, Laboratorio di Parassitologia, ISTISAN, Archivio Centrale dello Stato, Rome; Paul Russell, Introduction, 6.

8. Alberto Missiroli to the Gabinetto del Ministro Ministero dell'Interno, Rome, August 20, 1943; Alberto Missiroli to the Direttore dell'Ufficio d'Igiene del Governatorato di Roma,

November 2, 1943; Alberto Missiroli to the Direttore Generale dell'ISS, November 29, 1943; Alberto Missiroli to the Sottosegretario di Stato, August 24, 1944; "La Malaria nella Zona di Maccarese," January 24, 1947, all in busta 6, fasc. 19, Laboratorio di Parassitologia, ISTISAN, Archivio Centrale dello Stato, Rome; personal communication with Mario Coluzzi, son of Alberto Coluzzi, May 15, 2005.

9. Letter from Alberto Missiroli to the Sottosegretario di Stato, August 24, 1944; "La Malaria nella Zona di Maccarese," January 24, 1947.

10. L. W. Hackett, *Malaria in Europe: An Ecological Study* (London: Oxford University Press, 1937), 17; letter from George MacDonald to Missiroli, May 4, 1946, busta 6, fasc. 19, Laboratorio di Parassitologia, ISTISAN, Archivio Centrale dello Stato, Rome.

11. Letter from Missiroli, August 11, 1938, busta 6, fasc. 19, Laboratorio di Parassitologia, ISTISAN, Archivio Centrale dello Stato, Rome. See also Marion Hulverscheidt, "German Malariology Experiments with Humans, Supported by the DFG until 1945" (2003), http://www.geschichte.uni-freiburg.de/DFG-Geschichte/MedTagungAbstracts.htm, accessed June 15, 2005.

12. F. L. Soper, W. A. Davis, F. S. Markham, and L. A. Riehl, "Typhus Fever in Italy, 1943–1945, and Its Control with Louse Powder," *American Journal of Hygiene* 45, no. 3 (1947): 305–34, esp. 317, 320.

13. Edmund P. Russell, "The Strange Career of DDT: Experts, Federal Capacity, and Environmentalism in World War II," *Technology and Culture* 40, no. 4 (1999): 770–96, esp. 780; Paul Russell, Introduction, 9.

14. Paul Russell, Introduction, 5, 9.

15. Soper et al., "Typhus Fever in Italy, 1943–1945," 325.

16. Ibid.

17. Fred L. Soper, "Introduction of DDT to Italy, 1943–1945," *Rivista di Parassitologia* 20, no. 4 (1959): 403–9; Fred Soper to L. W. Hackett, January 20, 1959, Fred L. Soper Papers, U.S. National Library of Medicine; Blanche Armfield, *Organization and Administration in World War II* (Washington, DC: Office of the Surgeon General, Department of the Army, 1963), 292.

18. J. R. Busvine, "Eradicating the Mosquito," *Discovery* (March 1950): 85–89; Fred L. Soper, "Species Sanitation as Applied to the Eradication of (A) an Invading or (B) an Indigenous Species," in *The Proceedings of the Fourth International Congresses on Tropical Medicine and Malaria* (Washington, DC: Government Printing Office, 1948), 48; Col. D. E. Wright, "The Program of Insect Control on Crete" [1946], in United Nations Relief and Rehabilitation Administration records, Office of the Historian, Monographs Greece 22, Malaria & Sanitation, 1946 (S-1021–0034, 1945–1946, PAG 4/4.2.:34), United Nations Archives, New York City.

19. "Final Commentary: The Sardinina Project," Shell filmscript, received December 21, 1949 (Rockefeller Archive Center [hereafter RAC]), 1.2, 751, 16, 138), 3; John A. Logan, *The Sardinian Project: An Experiment in the Eradication of an Indigenous Malarious Vector* (Baltimore: Johns Hopkins University Press, 1953), 116; John Logan, diary entry for September 27, 1947 (RAC); "ERLAAS Monthly Report," April, 1949, 16 (RAC, 11.2, 700, 16, 135); "The Sardinian Project" (1948; 35 minutes), Shell Petroleum Company, Ltd.; "Battaglie di Pace," Penco Film (n.d.); "Adventure in Sardinia" (1950; 20 minutes), British Pathé in Association with the Nucleus Film Unit (RAC).

20. Edmund Russell, *War and Nature: Fighting Humans and Insects with Chemicals from World War I to Silent Spring* (New York: Cambridge University Press, 2001), 53–73.

21. Marston Bates, Preface, in Logan, *The Sardinian Project* (1953), x; quotation of ERLAAS sprayer taken from the Malaria Oral History Project, directed by Pier Luigi Cocco, Istituto di Medicina del Lavoro, University of Cagliari.

22. G. A. Canaperia and T. Patrissi, "La malaria in Italia nel periodo bellico e postbellico," *Rivista di Malariologia* 27 (February 1948): 1–28; letter from John Logan to Leach, June 16, 1947 (RAC, 1.1, 502, 1, 2).

23. Margaret Humphreys, "Kicking a Dying Dog: DDT and the Demise of Malaria in the American South, 1942–1950," *Isis* 87 (1996): 1–17.

24. Birori School, Sardinia, at http://www.macomer.net/scuolabirori/presentazione.html, accessed November 5, 2003. Interview quotations taken from Malaria Oral History Project, directed by Pier Luigi Cocco; Marcus Hall, "Today Sardinia, Tomorrow the World: Killing Mosquitoes," *BardPolitik: The Bard Journal of Global Affairs* 5 (Fall 2004), available online at www.bard.edu/bgia/bardpolitik/vol5/21-28.pdf, accessed April 15, 2009.

25. Letter from John Kerr to Fred Soper, June 6, 1946 (RAC, 1.2, 700, 12, 104).

26. John Logan, diary entry for November 19, 1948 (RAC).

27. Eugenia Tognotti, *Americani, Comunisti e Zanzare: Il piano di eradicazione della malaria in Sardegna tra scienza e politica negli anni della guerra fredda (1946–1950)* (Sassari: Collana Documenti e Opinioni, Editrice Democratica Sarda, 1995), 26, 47–48.

28. Marston Bates, in Logan, *The Sardinian Project*, xi.

29. Logan, *The Sardinian Project*, 389.

30. Dennis Mack Smith, *Italy: A Modern History* (Ann Arbor: University of Michigan Press, 1959), 494.

31. Marston Bates, "Retrospect," in *Man's Role in Changing the Face of the Earth*, ed. William L. Thomas et al. (Chicago: University of Chicago Press, 1956), 1140; Sardinia insect survey furnished by Dr. Carlo Contini, Cagliari, Italy.

32. Information about DDT poisoning of Sardinia's mullet stocks was furnished by Dr. Carlo Contini in a personal communication, April 2004.

CHAPTER 7

BIRDS ON THE HOME FRONT

Wildlife Conservation in the Western United States during World War II

Robert Wilson

The environmental consequences of war extend far beyond the battlefield. The pursuit of resources to sustain lengthy conflicts has often led to deforestation, soil erosion, and disruption of riverine environments as countries mobilized for war. Many of the contributors to this book argue that a nation's success or failure in war often depends on its ability to harness natural resources for military ends. An awareness that environmental change as a result of warfare can occur outside the areas of direct conflict is particularly important in relation to the United States, which was largely spared martial combat within its borders throughout the twentieth century.[1] Even though no battles were fought in the continental United States, the increased demand for natural resources during wartime is a vivid indication of the environmental consequences of sustaining military campaigns overseas.[2]

The connection between wildlife and war might appear strained, given that by the twentieth century most so-called game animals were no longer hunted commercially in the United States. Wildlife certainly played a significant role in earlier conflicts. In the nineteenth century, for example, the U.S. Army relied on bison and deer to feed troops engaged in military campaigns against Native Americans.[3] Later it attempted to exterminate the bison in parts of the American Midwest in order to undermine the resistance of bison-dependent Native American communities.[4] Yet by the beginning of World War I, game animals no longer constituted a major food source for most Americans; so, unlike natural resources such as timber and grain, government officials did not regard them as critical to the war effort.

While game animals may have been exempt from military uses, they, along with other wildlife, were directly and indirectly affected by wartime management and production of multiple natural resources. Timber harvesting in the Pacific Northwest modified the habitats of species dependent on mature forest stands. The federal government identified spruce wood as vital to airplane construction during World War I, which led to selective logging of the species in lowland regions of Washington State's Olympic Peninsula. Likewise, both world wars proved a boon to western mining operations, which supplied copper and other minerals crucial to military production.[5]

These wars drastically impacted conservation programs in the western United States. Federal land agencies, such as the Forest Service and the National Park Service, suffered severe cutbacks in funding and personnel. In World War I the National Park Service yielded to pressure from ranchers who claimed that permitting their herds to forage on park lands would help to provide beef for U.S. troops and allowed livestock grazing in California's Sequoia National Park. Once again in the Second World War, timber companies sought access to Sitka spruce in the Olympic National Park. The Park Service agreed to some logging but successfully resisted postwar attempts to reduce the size of the park.[6] Mobilization offered the perfect justification for reducing or eliminating environmental protections in the name of national security.

Arguably, World War II had a greater effect on the U.S. Fish and Wildlife Service (FWS)—known until 1940 as the U.S. Bureau of Biological Survey—than any other federal land-management agency. The agency was responsible for overseeing enforcement of federal wildlife laws, particularly those pertaining to migratory birds; for managing national wildlife refuges; and, most notoriously, for running a predator-eradication program in the western United States.[7] Poorly funded and staffed, the bureau languished for the first three decades of the twentieth century. In the early 1930s some conservationists called for its abolition, since it had proved unable to stop the decline in waterfowl populations. After the election of President Franklin D. Roosevelt in 1932 and the enactment of his New Deal conservation programs, the bureau's fortunes began to improve. Federal officials appointed by President Roosevelt, such as Secretary of Agriculture Henry Wallace and Secretary of the Interior Harold Ickes, sought a greater role for government agencies in conservation. Using funds and labor from the federally sponsored Civilian Conservation Corps (CCC), the Bureau of Biological Survey established dozens of new refuges along avian migration routes and initiated programs to rehabilitate damaged marshes. These programs signaled a profound shift in the government's role in wildlife protection. Long the handmaiden of farmers and irrigators, who often destroyed wildlife habitat, the government finally demonstrated leadership in stemming the decline of game-bird species, such as ducks and geese.

World War II put an end to most of these programs. Mobilization for the war placed a multitude of pressures on what was now the Fish and Wildlife Service and threatened to undermine the foundation of the conservation program that FWS personnel had so enthusiastically cultivated in the 1930s. This chapter analyzes the Fish and Wildlife Service's efforts to conserve migratory birds in the western United States during the war years. It also explores the consequences of technologies originally developed and used in a military context on the management of bird refuges in the years immediately after the war. I argue that the war's impact on migratory birds and on the agency that managed them was not entirely negative. Wars often cause *collateral damage* to forests, rivers, and coastlines, but the reduction of hunting pressures in the continental United States while the country was at war may also have led to *collateral productivity*. Duck and geese populations increased in the early 1940s, partially because fuel rationing and restrictions on ammunition made it difficult for sportsmen to travel to refuges and hunt waterfowl. To tease out these connections, I will briefly discuss specific ecological conditions and avian conservation programs in the western United States (especially in Oregon and California) in the late 1930s, the management of migratory birds and refuges between 1941 and 1945, and the war's lasting influence on bird management in the region.

Western Migratory Birds, Wildlife Refuges, and the New Deal

In the early twentieth century habitats for migratory birds had diminished considerably in Washington, Oregon, and California. A time-lapse film showing the changing map of wetlands in these states between 1890 and 1935 would illustrate the dramatic reduction and in many cases complete disappearance of marshes, vernal pools, and riparian vegetation. Over this period irrigation projects and development along rivers destroyed huge amounts of wetland habitat in the far west. Ninety percent of California's wetlands vanished, a greater percentage of wetland loss than any other state in the country.[8] Waterfowl and other migratory birds endured this onslaught of their habitat, though in greatly diminished numbers. In most places, few people opposed the draining of wetlands, and the government actively supported such projects.

Conservationists also took preliminary measures toward establishing sanctuaries that prohibited hunting. Early efforts focused on the Klamath and Malheur river basins of central Oregon, two staging areas used by birds while en route to wintering or breeding grounds. The Klamath Basin was particularly important since over 80 percent of waterfowl along the Pacific Flyway funneled into the area during their autumn migrations. These marshes were oases for migratory birds flying over the high desert. Although few birds wintered in these areas, most waterfowl stopped in them to rest and feed during the journey to their main wintering areas in California and Mexico. Every autumn as many as seven million waterfowl rested in these areas, feeding on aquatic plants or flying overhead, and creating, according to some observers, one of the great wildlife spectacles in North America.[9]

In the early twentieth century, several U.S. presidents established four migratory bird refuges in the Klamath Basin by executive order and placed them under the jurisdiction of the Bureau of Biological Survey.[10] The agency prohibited hunting on most of these refuges, yet it did little to ensure long-term protection of wetland habitats. The Klamath Basin was also the site of the Klamath Irrigation Project, one of the first projects undertaken by the U.S. Bureau of Reclamation. Beginning in 1905, the Bureau of Reclamation built dams, dikes, and canals to drain the basin's wetlands and redirect water to newly established homesteads. The bird sanctuaries within the basin were little more than nominal refuges; the Bureau of Reclamation regarded irrigation as the proper use of basin waters. Despite complaints from the Bureau of Biological Survey and bird conservation groups, the Bureau of Reclamation destroyed most of the Klamath Basin marshes. Much of the land reclaimed in this process proved unfit even for farming. By the late 1920s, Lower Klamath Lake—once a patchwork of open water and marsh plants covering 100,000 acres—had become a desert waste. When government ornithologist Frederick Lincoln visited the basin in 1935, the sight of what remained of the refuge disgusted him. "It doesn't even support a good crop of weeds.... A jack-rabbit would starve on it."[11]

In the later 1930s members of the Civilian Conservation Corps provided the labor for engineering projects that would allow the Bureau of Biological Survey to manage water on the refuges. President Roosevelt also pressured the Bureau of Reclamation to develop solutions to the water shortages on the Klamath Basin refuges. More importantly, congressional appropriations and revenue from federal waterfowl-hunting permits enabled the Bureau of Biological Survey to buy sites for refuges along such major migration routes as the Pacific Flyway (map 7.1).

The Central Valley of California was an equally important habitat for migratory birds. Nearly 60 percent of Pacific Flyway birds wintered there. By the 1930s, however, the valley's wetlands were rapidly disappearing. To deal with this habitat loss, the Bureau of Biological Survey established the 10,000-acre Sacramento National Wildlife Refuge in the northern half of the valley in 1936. The agency constructed ponds and created the conditions for marshes on the property to support the thousands of ducks and geese that wintered in the region. The refuge was envisioned as one unit in a constellation of refuges that the Bureau of Biological Survey and the California Department of Fish and Game would create amid the irrigated farmland in the valley. With the commitment of conservationists and the backing of the federal government, the goal of constructing a refuge network to sustain the continent's migratory bird populations seemed within reach.

The beginning of World War II placed this goal in question. Federal funds and labor disappeared as quickly as they had emerged in the mid-1930s. By early 1942 the CCC had disbanded, and many Fish and Wildlife Service personnel had left the agency to enter the armed forces.[12] Wildlife conservation was no longer a priority of the Roosevelt administration, which was now

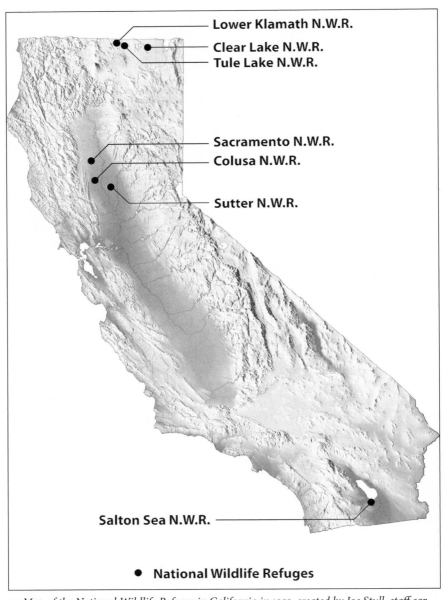

7.1 Map of the National Wildlife Refuges in California in 1955, created by Joe Stull, staff cartographer for the Department of Geography at Syracuse University.

focused on fighting a prolonged conflict against Germany, Japan, and Italy. World War II strained federal wildlife conservation in unexpected ways and, as I hope to show here, created an enduring legacy.

Wartime Conservation of Migratory Birds in California's Central Valley and Oregon and California's Klamath Basin

The Sacramento National Wildlife Refuge was intended to provide migratory waterfowl a sanctuary from overhunting. To attract larger numbers of waterfowl as well as other types of bird species, the FWS built ponds and developed marshes. Even though the service had spent most of the late 1930s constructing these ponds with CCC labor, by 1942 it had developed only 1,800 acres. As waterfowl populations recovered from their collapse during the previous decade due to improved conditions in the birds' breeding grounds, the Sacramento Refuge had to accommodate more and more ducks and geese.[13] It became apparent that the Sacramento Refuge and other federal refuges in the valley could not provide enough aquatic vegetation or cultivated crops for the birds to eat. The lack of an adequate food supply on the refuges prompted migratory ducks and geese to search for food on private farms nearby.

By the early 1940s, migratory waterfowl feeding in local rice fields were becoming a serious problem for farmers in California's Sacramento Valley. The rising number of waterfowl along the Pacific Flyway contributed to the problem, but the problem was exacerbated by changes in the surrounding landscape that were largely brought on by the booming rice market during the war.[14] In the rice-growing region of the Sacramento Valley, the number of acres in rice production had fallen from a high of 162,000 in 1920 to 88,000 in the 1930s. Federal officials had encouraged farmers to grow less rice throughout most of the depression years, but by 1942 rice acreage increased dramatically, reaching 192,000 acres. To satisfy demand, farmers began cultivating marginal areas. Prices for rice were high, but growing it under the conditions imposed by the war put enormous pressures on farmers.

Increased rice production in the Sacramento Valley had a mixed impact on migratory waterfowl. For some duck species, such as mallards, pintails, and wood ducks, and for all of the geese that wintered in California, rice served as a vital food source in a part of the flyway that had once supported extensive wetlands. However, farmers harassed birds that fed in rice fields with shotguns and flares, sometimes even killing them, an illegal activity but one that the FWS and California Department of Fish and Game found difficult to stop.[15]

Many rice-growing techniques contributed to the problem. Farmers flooded their fields early in the spring and drained them before harvesting in the fall. In their rush to increase production, farmers often chose not to level their fields each year, which was an expensive practice made even more so by wartime shortages of both labor and farm machinery. On the margins of rice fields, small areas of open water formed between the paddies and the small dikes (known as "checks") that surrounded them. Ducks were able to land on the shallow water in the paddies and begin feeding on the rice. As they fed, they created more open water in their wake, which attracted yet more waterfowl. Farmers and FWS biologists noticed that wood ducks, despite persistent harassment, often refused to move. Even worse for farmers, wood ducks usually left the refuge early in the morning to feed in the rice fields, a feeding pattern that attracted other ducks to the rice paddies.[16]

The numerous private hunting clubs throughout the Sacramento Valley did little to improve the situation. One such cluster combined seven gun clubs located around the Sacramento Refuge.[17] Most of these clubs were leased from rice farmers. During the hunting season, club operators partially flooded their land in order to entice waterfowl, but game regulations prohibited the dispersion of food for birds by gun-club owners. Since the temporary ponds created by gun clubs lacked the aquatic plants found in the marshes and riparian areas of the Sacramento Valley, the birds attracted to them found little or nothing to eat. Most clubs were either surrounded by rice fields or were situated near them; some clubs even grew rice.[18]

Rice farmers grew resentful of the FWS and the game policies it enforced. "No attempt should be made to belittle the supreme contempt with which a duck's life is considered in Colusa and adjacent counties," wrote one FWS official in California. "When conditions come to the point where some rice farmers studiously hunt for and destroy all waterfowl nests found in their fields, one has reached a climax in anti-conservation."[19] Farmers complained that they were providing feed for waterfowl that were then hunted merely for sport, mostly by wealthy gun-club members. There was truth in their accusations. Hunting was forbidden in the Sacramento Refuge, and as of 1942 the state and federal governments had established no public hunting grounds in the Central Valley. Unless they belonged to gun clubs, less-affluent sport hunters had no place to hunt ducks and geese legally. FWS officials feared that without significant action on these issues, the entire waterfowl conservation program in California might collapse.

To save the program in the Central Valley, the FWS needed to address the concerns of farmers and to develop support for (or at least acceptance of) more federal and state refuges. By allowing seasonal duck hunting on future refuges, the FWS hoped to improve its relations with the public. Agency officials described the refuge program as so "closely tied in with the proposal of a land-use policy, and with the operation of feed units, that it can not be longer avoided. It would be the most popular move possible in the Sacramento Valley."[20]

Beginning in the mid-1940s, the FWS developed a program to lure waterfowl out of private farms and into the refuges. The FWS had grown crops as feed for migratory waterfowl on the Tule Lake and Sacramento refuges during the late 1930s, but the outbreak of war prevented the agency from expanding the program. Agency officials debated whether FWS personnel should grow the crops or whether it should hire farmers to sharecrop on the refuge. Under the latter arrangement, farmers would harvest part of the crop for themselves and leave the remainder as food for the birds. E. E. Horn, the chief FWS biologist in California, doubted that many farmers would agree to sharecrop since many of them were struggling to cope with waterfowl on their own lands. Ducks and geese might wipe out the farmers' share of the grain as well as the grain designated for the birds. Horn predicted that the FWS would have to run and staff the program itself if it hoped to raise sufficient feed.[21]

In 1944 the FWS leased 800 acres near the town of Colusa and 420 acres in the Sutter Bypass, on which the service cultivated rice.[22] The following year the FWS purchased these properties (and other land nearby) to create the Colusa and Sutter National Wildlife Refuges. The FWS managed the Sacramento Refuge as a mixture of crops, ponds, and natural vegetation. In 1945 it planted 335 acres of barley, 475 acres of rice, and flooded 1,500 acres of ponds and marshes. In 1948 the U.S. Congress passed the Lea Act, which provided federal funding for the purchase of land that would be used to create waterfowl feeding grounds throughout the agricultural regions of California. In a move popular with sportsmen, it also opened these new refuges to hunting.

The goal was to create small feeding areas throughout the agricultural regions of California to disperse waterfowl. The California Department of Fish and Game also developed refuges (called Waterfowl Management Areas) to raise food for waterfowl. Most of these federal and state refuges were between 2,500 and 8,000 acres, and, with the exception of the Sacramento Refuge, all were open to hunters. Collectively they amounted to a string of islands for waterfowl within the agricultural landscape of the Central Valley.[23]

To divert waterfowl away from farmland, refuge staff and farmers experimented with a number of methods. In addition to harassing birds with shotguns and flares, farmers in the Klamath Basin stationed watchmen in the fields at night to prevent ducks and geese from devouring crops. To aid the farmers, the FWS developed mortars that projected small explosives into the air above flocks of feeding waterfowl. More often than not, these explosions merely prompted the birds to move from one farm to another. Over time, the birds grew accustomed to the harassment and stopped moving very far. In the Klamath Basin and Imperial Valley of California, farmers and refuge officials were moderately successful using antiaircraft searchlights to prevent waterfowl from landing in their fields at night. When farmers were vigilant, they could keep birds from destroying at least part of their crops; but if they let their guard down, ducks and geese could wipe out acres of rice, barley, or lettuce in a matter of hours.[24]

Aerial herding was the most successful method for redirecting birds into refuges. Using surplus aircraft acquired from the military during World War II, FWS pilots swooped in above flocks of waterfowl in grain fields, sometimes dropping grenades among the birds to frighten them into the air. Once the birds were aloft, the pilots herded the ducks and geese like cattle into the refuges. Even with aerial herding, depredations remained a serious problem after the war. The numbers of migratory waterfowl were still high at the time. In the Klamath Basin, for instance, up to four million waterfowl rested and fed on Tule Lake Refuge at the peak of the fall migration. It was nearly inevitable that the birds would come into conflict with farmers who were homesteading land nearby or leasing land on the refuges.[25]

War Relocation Camps and Wildlife Refuges

As with federal refuges elsewhere in the nation, labor and funding for the FWS decreased significantly in the war years. The relocation of the FWS headquarters from Washington, DC, to Chicago in 1942 further hampered the agency's ability to manage refuges effectively and vividly symbolized the marginalization of conservation during the war years.[26] The FWS faced other problems, too. In April 1942, the newly formed War Relocation Authority (WRA) began constructing an internment camp near the southeast boundary of the Tule Lake Refuge, one of many that the federal government built during World War II to intern Japanese Americans. The largest internment camp, it eventually housed over 18,000 people. There is little indication that the WRA or the Department of the Interior consulted with the FWS before building the camp near the refuge. J. Clark Salyer II, director of the FWS's Division of Refuges, was furious about the placement of the camp, saying that the agreement struck between the WRA and the Bureau of Reclamation reduced the role of the FWS to that "of an innocent bystander."[27] Even though the camp was not built on the refuge, Salyer worried that hunters might be barred from the public shooting grounds on the refuge located nearest the facility. There was also the possibility that the War Relocation Authority might build additions to the internment camp on the refuge itself.[28]

According to the Bureau of Reclamation, the land on which the camp was situated was not private property but land leased to farmers on the Klamath Project. The agency did not hesitate to terminate the leases since the WRA claimed these internment camps were needed for national security. To the WRA, this was a suitable location for many reasons: it was relatively isolated, it was far from the Pacific Coast, and the federal government already owned it. The WRA could construct a camp on the property without having to purchase it from private owners. The agreement between the Bureau of Reclamation and the WRA showed a near-total disregard for the refuge and for the wildlife that depended on it.[29]

While the WRA did not use land within the Tule Lake Refuge for the camp, it did start a farming program on the refuge. Salyer worried that FWS could not protect waterfowl in the areas where internees grew crops, and that the birds would devour many of the grains and vegetables they planted. He accused the internees of endangering flocks of waterfowl by driving trucks through them and of harvesting ducks and geese illegally, using fishhooks baited with food to capture the birds. Salyer also felt that the WRA wasted many of the vegetables that the internees grew on the refuge by allowing potatoes and lettuce to freeze in the fields while the WRA delivered boxcars of food to the Tule Lake Camp.

In his letters to Ira Gabrielson and other FWS officials, Salyer revealed that he shared the same distrustful attitude toward the Japanese American internees expressed by many other Americans during the war. He considered the internment camp a menace to the migrating waterfowl and likened the Tule Lake Refuge to a group of American soldiers held captive by the Japanese military. He also claimed that the WRA had little control over the internees, who could easily obtain passes to leave the camp; and while working in the fields, the internees pestered waterfowl and made bird management difficult. In Salyer's view, the credibility of the FWS would suffer among sportsmen if the internees were not more closely supervised.[30]

At the same time, the agency also sought to use Japanese American internees as laborers on refuge conservation projects. With the demise of the CCC shortly after the United States entered the war, the FWS was forced to scramble for manpower to complete construction projects begun in the 1930s. Japanese American internees at WRA camps seemed like a convenient substitute for the departing CCC men. In summer of 1942, Salyer urged the WRA to allow internees from the Tule Lake camp to work on refuge projects. Salyer also hoped to use internees to help develop a proposed refuge in Owens Valley, California, near the site of the Manzanar War Relocation Center. The WRA refused both requests, saying that internees were needed to work on camp farms and other projects directly associated with the relocation centers. Although unsuccessful, the FWS attempt to use internees for its own conservation objectives demonstrated a callous disregard for the plight of Japanese Americans during the war.[31]

Despite Salyer's concerns, the internment camp and its farming program caused no more damage than similar farming operations on the Tule Lake Refuge, and Klamath Project farmers probably would have leased that land if the War Relocation Authority had not done so. It did, however, show the unwillingness of other federal agencies to consult with the FWS before they took action that might affect federal refuges. Over the 1930s the FWS had established itself as a partner in the use of the basin's water and as an advocate for migratory waterfowl. Once the war began, the War Relocation Authority and the Bureau of Reclamation developed a farming program for the Tule Lake internment camp despite the criticisms of the FWS. This controversy showed how quickly the government could jettison conservation objectives in wartime.

In the Shadow of the War

After World War II, the FWS tightened the links between federal refuges and the irrigation system set up by the Bureau of Reclamation and irrigation districts to support agriculture in far-western states. For the refuges to function, the FWS needed plentiful water to fill ponds and to irrigate the crops it grew for waterfowl. Networks of canals and drains now brought water to irrigated farms, and to obtain it, the FWS integrated the refuges into this system. In effect, the refuges became another unit in the irrigated landscape created by reclamation.

There were other links between modern agriculture and refuges as well. In the 1940s and 1950s the FWS adopted the same technologies to manage its refuges that farmers in California and Oregon used. Refuge officials relied on chemical insecticides and herbicides developed during World War II to safeguard their crops and boost production. Like the commercial wheat and rice that surrounded the refuges, the monocrops grown on them were susceptible to the insect pests and weeds that passed freely between refuges and surrounding private farms. The FWS found itself in a challenging predicament. Reclamation had destroyed most of the waterfowl habitat in the wintering range, but in order to save the birds, the FWS felt it had little choice but to transform large portions of the refuges into mirrors of the agricultural landscape.

Pests found in irrigated fields also thrived on the refuges. Often development on the refuges promoted the spread of unwanted organisms. Weeds in particular proved a troubling problem. The miles of dikes constructed in the Sacramento and Klamath Basin refuges proved ideal environments for weed dispersal. Cattails often clogged the canals that distributed water throughout the refuges. The growth and spread of such plants were more than nuisances; together they undermined the agency's capacity to manage the refuges. During the 1930s the Bureau of Biological Survey used machinery such as draglines to remove cattails from irrigation canals, although this work was sometimes done by hand—a labor-intensive and time-consuming process, feasible only when workers from the CCC were available.[32]

The service had already experimented with chemical agents. In the 1920s, the Bureau of Biological Survey, in conjunction with the U.S. Chemical Warfare Service, tested such chemicals as mustard gas, chlorine, and chloropierin (all of which were used in combat during World War I) on small birds that farmers considered pests.[33] At that time the Biological Survey was also studying the utility of birds in controlling insects. The agency hoped that use of these "war gases" could prevent the destruction of crops by blackbirds in California's Imperial Valley, but researchers found that the birds escaped from the poisons before they could take effect.[34]

In the late 1940s the FWS routinely sprayed the herbicide 2,4-D on refuges in the Sacramento Valley, Klamath Basin, and the Harney Basin of eastern Oregon. This herbicide was a common tool used by farmers on private lands. On the Sacramento Refuge, for instance, managers started applying herbicides in 1947 to waterways and plots of land on an experimental basis.[35] Many of the new structures built within the refuges—especially the miles of dikes constructed by the CCC and FWS—encouraged the spread of weeds such as Russian thistle and mustard. To curb this spread, the FWS also sprayed the dikes with 2,4-D.³⁶ But even when spraying killed the weeds, they often grew back very quickly. Herbicides had little effect on cattails, one of the most virulent weeds on the refuge. Wildlife officials also worried that herbicides might kill plants that were a food source for waterfowl. Later in the 1950s the FWS contracted with the B. F. Goodrich Chemical Company to test other herbicides, including VL-600, on refuge lands.³⁷

Much of the land within the refuges was marginal and already infested with weeds. Warren S. Bourn, a researcher in the FWS's Section of Habitat Improvement, argued in a 1949 memo that herbicides were necessary in order to bring these lands into crop production for waterfowl feed and to limit the spread of weeds from refuges to private farms. He recognized that such spraying might spark controversy and stated that "in no case is the use of 2,4-D or similar chemicals on refuge areas to be publicized." He feared that farmers could blame the agency for crop damage due to botched spraying whether it was justified or not. "The loss of a crop from blight, insects or other causes may conceivably be blamed on 2,4-D use in the vicinity. Therefore, appropriate secrecy should attend its application on refuge areas."³⁸ Even at this early date, the FWS realized that new chemicals such as 2,4-D could spread beyond the refuges and lead to complaints against the agency.

To control insect pests, the FWS used the insecticide DDT, a choice that would have lasting consequences for the refuges and for the surrounding environment. Like the herbicide 2,4-D, DDT was one of a number of pest-fighting chemicals developed in the 1930s and applied by the U.S. military during World War II.³⁹ In the Pacific campaign, the military found the insecticide to be effective in controlling lice and in killing mosquitoes, which carried malaria. By the end of the war, farmers and municipalities were already beginning to spray DDT on rural farms and urban parks as part of a move to bring the war against insects from the battlefield back to the United States. Although DDT was the best known of the new insecticides, others belonging to a group of chemicals known as chlorinated hydrocarbons and organophosphates were also widely employed by farmers and eventually the FWS.⁴⁰

The FWS's eagerness to spray DDT on the refuges is surprising given that the agency was one of the first institutions to study the effects of this insecticide on birds and mammals. Analyses conducted in the late 1940s showed that when incorrectly applied, DDT could cause immediate death in some mammals and fish. Such studies suggested that problems with DDT were attributable to incorrect dosages rather than to the chemical per se. Investigators looked for unintended mortalities due to poisoning, not for deaths caused by long-term exposure to the chemical. Researchers at the agency's Patuxent Wildlife Research Center in Maryland discovered that DDT could affect the reproduction of birds exposed to it, even in very small quantities. Other scientists not affiliated with the FWS would later argue that the insecticide had environmental consequences far beyond its effect on insects targeted by spraying programs.⁴¹

Farmers adopted DDT and other chemical insecticides during the early 1950s, and the FWS soon followed their lead. The FWS considered the new insecticides a cheap and efficient way to control insects that harmed crops. The agency believed that applying the chemicals would help to boost production of grain for waterfowl. However, pressure to use the chemicals also came from farmers, state agricultural agencies, and the U.S. Department of Agriculture, all of whom saw the refuges as potential havens for insect pests. On a number of occasions, regional insect-control programs affected the refuges. In 1953 an outbreak of rice-leaf miners spread through the rice-growing districts in the Sacramento Valley. To bring the outbreak under control, farmers spent $1.2 million to spray more than 200,000 acres with dieldrin, a chlorinated-hydrocarbon pesticide similar to DDT. Farmers sprayed rice fields at the refuge edge, and insecticide mists from these applications drifted onto refuge fields and waterways.[42] The FWS also signed agreements with the Department of Agriculture to deal with insect problems in agricultural areas. In 1956 the FWS and the Plant Control Branch of the Department of Agriculture agreed to use DDT and other insecticides to control a grasshopper infestation on the Tule Lake Refuge. The agencies worried that if left unchecked, the insects would cause "serious damage to such croplands as well as to adjacent crops on lands leased to private individuals."[43]

Although in the 1950s scientists did not understand the long-term consequences of exposure to DDT and other insecticides, there were already clear signs that the chemicals could have serious effects on species not targeted by the spraying. Farmers' use of insecticides to kill rice-leaf miners in the Sacramento Valley also killed various wading birds, such as snowy egrets, black-crowned night herons, and great blue herons on the Sacramento Refuge; refuge staff estimated that up to 10 percent of these bird populations were lost, as well as some ducks and upland birds.[44] Richard E. Griffith, chief of the FWS's Section of Habitat Improvement, wrote that this was probably unavoidable: "such losses are a calculated risk in the treatment for crop protection. In the case of the refuge rice fields, we will have to accept some slight losses if we are to produce the large amount of grain needed for waterfowl management during the critical harvest season."[45]

Even worse die-offs occurred on the Tule Lake Refuge. In 1960 refuge personnel noticed hundreds of dead grebes, pelicans, gulls, and egrets—all fish-eating birds—within the Lower Klamath and Tule Lake refuges. Analysis of the carcasses revealed that many contained high levels of the chlorinated-hydrocarbon insecticides toxaphene, DDT, and DDD. Of the nearly 360 dead birds collected by refuge staff, over 200 were white pelicans. FWS scientists suspected that the greater mortality of white pelicans was due to the type of fish they ate. Larger fish tested in the area had much higher concentrations of pesticides than smaller fish. Even more startling was the concentration of pesticides found in the developing eggs of one of the cormorants tested. The female had 2.6 parts per million of the pesticides, but the developing eggs

Birds on the Home Front

contained 44 parts per million—nearly seventeen times more than an adult cormorant.[46]

Such a large die-off of birds on a wildlife refuge was bound to attract attention. Rachel Carson, a former editor and scientist for the FWS, discussed the death of the birds in her indictment of pesticides, *Silent Spring*. In her rendering, the birds' deaths were the result of insecticides finding their way into canals that drained wastewater from private farms into the refuges, not from the application of pesticides by the FWS.[47] The failure to mention FWS spraying is puzzling. As a former employee of the FWS, Carson had authored a series of educational booklets called *Conservation in Action* about important federal refuges in the country, and she must have been aware that the FWS used insecticides and herbicides. Yet throughout *Silent Spring*, she portrays the FWS as the investigator of pesticide effects on wildlife and not as a pesticide user.[48]

While the FWS was a leading investigator of the ecological effects of insecticides and herbicides in the postwar period, it was also a frequent user of DDT, 2,4-D and other pesticides. The spraying of pesticides and herbicides was an important way for the FWS to achieve its conservation objectives. Insecticides and herbicides were cheap and effective tools for eliminating pests that could damage crops intended as feed for migratory waterfowl. The "war against insects" begun during World War II became a national conservation strategy in the postwar years.

Conclusion

World War II had uneven effects on the programs and practices of federal land-management agencies in the United States. The National Park Service and the Forest Service were pressured to ease environmental regulations in order to support the provision of natural resources for the war effort. Individuals within and outside government fought vigorously to prevent the weakening of such regulations with varying degrees of success. Their stories are cautionary tales that demonstrate the ease with which hard-won conservation victories can be undermined in wartime.

The war certainly helped to rejuvenate the rice industry in California, which in turn created more food sources for migrating waterfowl in their wintering range. Ammunition restrictions and fuel rationing made it more difficult for hunters to practice their sport during the war. National wildlife refuges, like national parks, witnessed enormous increases in visitors after the war. The long postwar boom led to a rise in outdoor recreation, enabling thousands to practice bird hunting. Some of these economic and social changes might have occurred even if the war had never taken place.

Construction of the Tule Lake internment camp next to a wildlife refuge in the Klamath Basin certainly would not have happened without the war. But as strange as this was, the use of the refuge by the War Relocation Authority and its disregard of the FWS's objections was in many ways a continuation of preexisting trends. The FWS had long been the weaker agency in the basin, and the Bureau of Reclamation had consistently dismissed its concerns. The construction of the internment camp and the opening of the refuges to farming was merely a bizarre chapter in the story of other federal agencies marginalizing the FWS.

For environmental historians the narrative of wildlife conservation in the western United States during World War II raises questions of causality. Since military combat did not occur in the region, there was no direct collateral damage on forests or fields due to conflict. However, the extent of the U.S. government's mobilization and prosecution of this multiyear war had far-reaching effects on the environment within the continental U.S. and on the agencies charged with managing wildlife. While the war was certainly not the only factor affecting migratory waterfowl during the early 1940s, it played a critical role in transforming the wetland habitats on which they depended. This study shows the powerful, and sometimes contradictory, ways that war can influence the economic and political processes that cause environmental change.

Notes

1. Obvious exceptions to this were the Japanese attacks on Hawai'i and the Aleutian Islands of Alaska during World War II.

2. Richard P. Tucker, "The World Wars and the Globalization of Timber Cutting," in *Natural Enemy, Natural Ally: Toward an Environmental History of War*, ed. Richard P. Tucker and Edmund Russell (Corvallis: Oregon State University Press, 2004), 110–41.

3. William Cronon, *Nature's Metropolis: Chicago and the Great West* (New York: W. W. Norton, 1991), 216.

4. This was only one of many factors that contributed to the precipitous decline of bison during this time. For a thorough analysis of the reasons for the near extermination of the bison, see Andrew C. Isenberg, *The Destruction of the Bison: An Environmental History, 1750–1920* (Cambridge: Cambridge University Press, 2000).

5. Gerald D. Nash examines some of the social and economic effects of World War II on the western United States in *The American West Transformed: The Impact of the Second World War* (Lincoln: University of Nebraska Press, 1985) and *World War II and the West: Reshaping the Economy* (Lincoln: University of Nebraska Press, 1990). See also Richard White, *"It's Your Misfortune and None of My Own": A New History of the American West* (Norman: University of Oklahoma Press, 1991), 496–533.

6. Richard West Sellars, *Preserving Nature in the National Parks: A History* (New Haven, Conn.: Yale University Press, 1997), 85, 152–53.

7. An administrative history of the U.S. Fish and Wildlife Service has yet to be written, but for a brief overview, see Mark Madison's entry on that agency in *The Encyclopedia of World Environmental History*, ed. Shepard Krech III, John R. McNeill, and Carolyn Merchant (New York: Routledge, 2004). On predator control, see Thomas R. Dunlap, *Saving America's Wildlife* (Princeton, NJ: Princeton University Press, 1988), 48–61; and Donald Worster, *Nature's Economy: A History of Ecological Ideas*, 2nd ed. (Cambridge: Cambridge University Press, 1994).

8. Oregon lost 38 percent, Washington 31 percent. William J. Mitsch and James G. Gosselink, *Wetlands,* 3rd ed. (New York: John Wiley and Sons, 2000), 84. These figures represent wetland loss up to 1990, but much of it occurred before the 1940s. See also W. E. Frayer, Dennis D. Peters, and H. Ross Pywell, *Wetlands of the California Central Valley: Status and Trends, 1939 to mid-1980s* (Portland, Ore.: United States Fish and Wildlife Service, 1989).

9. Tupper Ansel Blake, Madeleine Graham Blake, and William Kittredge, *Balancing Water: Restoring the Klamath Basin* (Berkeley: University of California Press, 2000), 1.

10. The four refuges instituted by presidential order are: Lower Klamath Lake Refuge (1908), Clear Lake Refuge (1911), Upper Klamath Lake Refuge (1928), and Tule Lake Refuge (1928).

11. Frederick C. Lincoln, "Western Field Trip, 9/7–10/20/35," Records of the Bureau of Sport Fisheries and Wildlife, Records of the Division of Wildlife Research, Office Files of Frederick C. Lincoln, 1917–60, Record Group 22, National Archives and Record Administration–College Park, Maryland (hereafter cited as RG 22, NARA–CP).

12. As mentioned at the beginning of this chapter, the Bureau of Biological Survey changed its name to the U.S. Fish and Wildlife Service in 1940.

13. Clinton H. Lostetter, "They've Got to Eat Someplace," in *Flyways: Pioneering Waterfowl Management in North America,* ed. A. S. Hawkins, R. C. Hanson, H. K. Nelson, and H. M. Reeves. (Washington, D.C.: U.S. Department of the Interior, Fish and Wildlife Service, 1984), 461–62.

14. Johnson A. Neff, Peter J. Van Huizen, and James C. Savage, "Rice and Ducks in the Sacramento Valley of California, Season of 1942," January 28, 1943; and California Farm Bureau Federation, "Statement of Position of California Farm Bureau Federation on Crop Damage Caused by Migratory Wild Fowl, May 27, 1943," Research Reports, 1912–51, Records of the Branch of Wildlife Research, RG 22, NARA–CP, 3.

15. E. E. Horn, "The Duck Problem," January 6, 1944, Office Files of Frederick C. Lincoln, 1917–60, Records of the Bureau of Sport Fisheries and Wildlife, Records of the Branch of Wildlife Research, RG 22, NARA–CP.

16. Neff, "Rice and Ducks in the Sacramento Valley," 8–11.

17. Ibid., 13–14.

18. Ibid.

19. Neff, "Rice and Ducks in the Sacramento Valley," 16. Some farmers wanted the refuges abolished. See Narrative Report–Sacramento National Wildlife Refuge (hereafter NR–SNWR), September–December, 1947, 22–23, Sacramento National Wildlife Refuge-Headquarters (hereafter SNWR–HQ).

20. Neff, "Rice and Ducks in the Sacramento Valley," 46–47. Conservationists and game officials had long worried about restricting public access to hunting lands. For instance, see Aldo Leopold, *Game Management* (New York: Scribner's, 1933).

21. Horn, "The Duck Problem."

22. The Sutter Bypass was one of the main canals in the Sacramento Flood-Control Project. Except in times of flood, this canal was mostly dry. To run its farming program, the project diverted water from irrigation districts.

23. "Waterfowl Pose California Problem," Outdoor California, September 1935, 1, 3; J. Clark Salyer II and Francis G. Gillet, "Federal Refuges," in *Waterfowl Tomorrow,* ed. Joseph Linduska (Washington, D.C.: U.S. Department of the Interior, Bureau of Sport Fisheries and

Wildlife, Fish and Wildlife Service, 1964), 504; and Albert M. Day, *North American Waterfowl* (New York: Stackpole and Heck, 1949), 168, 173.

24. "Summary of Year's Activities," Narrative Report–Klamath Basin National Wildlife Refuges (hereafter NR–KBNWR), 1944; and "Annual Summary," NR–KBNWR, 1947, 4, Klamath Basin National Wildlife Refuges–Headquarters (hereafter KBNWR–HQ).

25. NR–KBNWR, Tule Lake National Wildlife Refuge, September–December 1955, 24, KBNWR–HQ.

26. Ira Gabrielson, "Memoirs of Ira M. Gabrielson (and What Others Have Said about Him), 1889–1977," unpublished manuscript, National Conservation Training Center, Shepherdstown, West Virginia, 298–300.

27. J. Clark Salyer II to Regional Director Leo L. Laythe, June 3, 1942, "Tule Lake, 1934–1944," Bureau of Biological Survey General Correspondence, 1890–1956, RG 22, NARA–CP.

28. Chas. E. Jackson, Acting Director, to William Finley, June 24, 1942, "Tule Lake, 1934–1944," Bureau of Biological Survey General Correspondence, 1890–1956, RG 22, NARA–CP.

29. See "Memorandum of Understanding between the Director of the War Relocation Authority and the Secretary of the Interior," undated (internal evidence suggests April 1942), "Tule Lake, 1934–1944," Bureau of Biological Survey General Correspondence, 1890–1956, RG 22, NARA–CP.

30. Memorandum, J. Clark Salyer II to Mr. Day, Mr. Chaney, and Mr. Gardner, October 23, 1942; Dillon to Leo Laythe, Telegram, November 13, 1942; and Memorandum, J. Clark Salyer II to Dr. Gabrielson, January 26, 1943, "Tule Lake, 1934–1944," Bureau of Biological Survey General Correspondence, 1890–1956, Box 152, RG 22, NARA–CP.

31. J. Clark Salyer, Chief, Division of Wildlife Refuges, to Leo Laythe, Regional Director, FWS, Portland, June 12, 1942, FF20, Box 2, J. Clark Salyer Papers, Denver Public Library, Western History/Genealogy Collection (hereafter JS–DPL); Paul T. Kreager, Regional Refuge Supervisor to Director, Fish and Wildlife Service, May 27, 1942, May 29, 1942, June 5, 1942, FF13, Box 11, JS–DPL; C. G. Fairchild, Refuge Manager, to Leo L. Laythe, Regional Director, June 23, 1942, FF13, Box 11, JS–DPL; A. C. Elmer, Assistant Chief, Division of Wildlife Refuges, to Leo L. Laythe, Regional Director, July 13, 1942, JS–DPL.

32. NR–SNWR, 1939, SNWR–HQ.

33. No doubt, this was part of the Chemical Warfare Service's attempts to find "peaceful" applications for gases used during World War I. See Edmund P. Russell III, "'Speaking of Annihilation': Mobilizing for War against Human and Insect Enemies," *Journal of American History* 82, no. 4 (1996): 1510–18.

34. E. R. Kalmbach, "Report on Experiments in the Use of War Gases as Bird Control Agencies, Conducted at the Edgewood, Md., Arsenal of the Chemical Warfare Service," War Gases as Bird Control Agencies [sic], Research Reports, 1912–51, Records of the Fish and Wildlife Service, 1922–61, Records of the Branch of Wildlife Research, RG 22, NARA–CP. The Biological Survey did poison other animals such as coyotes and wolves in the western United States. Matthew D. Evenden, "The Laborers of Nature: Economic Ornithology and the Role of Birds as Agents of Biological Pest Control in North American Agriculture, ca. 1880–1930," *Forest and Conservation History* 39 (October 1995): 172–83; Dunlap, *Saving America's Wildlife*, 38, 76, 112–14; and Worster, *Nature's Economy*, 263–64.

35. NR–SNWR, Sacramento National Wildlife Refuge, 1947, 13, SNWR–HQ.

36. NR–SNWR, Sacramento National Wildlife Refuge, 1951, 10; and NR–SNWR, Sacramento National Wildlife Refuge, 1953, 29, SNWR–HQ.

37. Rossalius C. Hanson, Pilot-Biologist, to Regional Director, Portland, OR, "Report on Use of Herbicides on Sacramento Refuge," December 28, 1950; and Warren S. Bourn to Regional Director, Portland, OR, "Procurement and Use of Herbicides," March 31, 1953, Sacramento, 1945–55, Habitat Improvement, Division of Wildlife Refuges, RG 22, NARA–CP; NR–SNWR, Sacramento National Wildlife Refuge, May–August 1953, 9, SNWR–HQ. For herbicide use on the Malheur National Wildlife Refuge, see Nancy Langston, *Where Land and Water Meet: A Western Landscape Transformed* (Seattle: University of Washington Press, 2003), 106–8 and figure 7.

38. Memorandum, Warren S. Bourn, "2,4-D and Its Uses," January 25, 1949, Control-General, Wildlife Refuges, 1946–55, Record Group 22, National Archives and Record Administration—Pacific Alaska Region, Seattle, Washington.

39. Edmund Russell III, *War and Nature: Fighting Humans and Insects with Chemicals from World War I to Silent Spring* (New York: Cambridge University Press, 2001), 86. On the development of insecticides during this period, see John H. Perkins, *Insects, Experts, and the Insecticide Crisis: The Quest for New Pest Management Strategies* (New York: Plenum Press, 1982), 3–22.

40. Russell, *War and Nature*, 95–118, 165–75.

41. The federal government supported research into DDT during the 1950s under the direction of the FWS; see Thomas R. Dunlap, *DDT: Scientists, Citizens, and Public Policy* (Princeton, NJ: Princeton University Press, 1981), 76–97.

42. Norris A. Bleyhl, "A History of the Production and Marketing of Rice in California" (PhD diss., University of Minnesota, 1955), 182–83; and NR–SNWR, Sacramento National Wildlife Refuge, May–August 1953, 5, 9, SNWR–HQ. On the characteristics of the pesticide dieldrin, see Shirley A. Briggs and the staff of Rachel Carson Council, *Basic Guide to Pesticides: Their Characteristics and Hazards* (Washington, DC: Hemisphere Publishing, 1992), 134, 212–13.

43. "Memorandum of Understanding between the U.S. Department of Interior, Fish and Wildlife Service, and the U.S. Department of Agriculture, Plant Pest Control Branch," July 16, 1956, NR–KBNWR, Tule Lake National Wildlife Refuge, September–December 1956, 13–14, KBNWR–HQ.

44. The FWS also characterized the loss of fish as "severe." However, refuge officials were unconcerned by this since "the kill was predominantly rough fish and the resultant loss to fishermen was considered not too important." The FWS did not participate in this spraying. NR–SNWR, Sacramento National Wildlife Refuge, May–August 1953, 2, 5, 9, SNWR–HQ.

45. Richard E. Griffith to Regional Director, Portland, OR, July 3, 1953, Sacramento, 1945–55, Habitat Improvement, Division of Wildlife Refuges, RG 22, NARA–CP.

46. The results of the laboratory analyses can be found in the Klamath Basin Refuge's report from the following year. See NR–KBNWR, Tule Lake National Wildlife Refuge, May–August 1961, 7–8, 16, KBNWR–HQ.

47. Rachel Carson, *Silent Spring* (Boston: Houghton Mifflin, 1962), 49–50.

48. Linda Lear discusses Carson's duties and experience working for the Fish and Wildlife Service in *Rachel Carson: Witness for Nature* (New York: Henry Holt, 1997), 95–96, 102, 106–11, 132–33, 138–46.

CHAPTER 8

CREATING THE NATURAL FORTRESS

Landscape, Resistance, and Memory in the Vercors, France

Chris Pearson

Under a blazing hot July sun, French Prime Minister JeanPierre Raffarin helicoptered into the village of Vassieux-en-Vercors to mark the sixtieth anniversary of the deaths of more than eight hundred resistance fighters (otherwise known as *maquisards*) and civilians killed in the German assault on the Vercors during the summer of 1944, an attack intended to crush the resistance activity that been germinating in the area since 1941. Raffarin had come, he said, "to pay the homage due to all the resisters . . . of this place, which is so emblematic of their values, their courage, and their historical role."[1] His choice of words was apt, as over the years the rocks, soil, and dramatic limestone cliffs of the Vercors have come to embody armed French resistance to the German occupation. Indeed, it was in the Vercors that France "made real war," in the words of General de Lattre de Tassigny, attempting to defend France's wartime record.[2] The imposing topography of the Vercors and its three hundred resistance memorials now combine to create a monumental landscape.[3] The sheer range of memorial sites—which includes traditional monuments and cemeteries, a ruined village, and a memorial center—attests to the historical complexity of this resistance and the ways in which it is remembered. Furthermore, its increasing institutionalization demonstrates the national importance of the Vercors; from 1994 onward, the *Site national historique de la résistance en Vercors* (SNHRV) has united these memorials under the auspices of the *Parc naturel régional du Vercors* (PNRV), an organization created in 1970 to preserve the cultural and natural heritage of this mountainous area of 186,000 hectares on the western fringe of the Alps.[4]

Introducing the Memorial Environment

Historians have traced the emotive history of resistance in the Vercors and its tortured memory in great detail.[5] The environment, however, does not feature prominently in these narratives. Where it does appear, it is treated unproblematically as a static backdrop for human activity and aspirations. Michael Pearson's approach is typical. He states that "this enormous natural citadel was designed for drama, yet strangely it was ignored by history until 1944," when it became the "setting of the most famous saga of the French Resistance."[6] Environmental determinism and a rigid, ahistorical conception of nature form the methodological backbone of these histories. Additionally, the ecological consequences of the war on the biota of the Vercors await thorough investigation. This myopia toward nature reflects a more general neglect of the environment in analyses of French memory of the Second World War, especially in comparison with the attention devoted to the commemorative landscapes and garden cemeteries associated with the First World War.[7] It is almost as if the later war and occupation are considered to have left no physical traces.[8]

The environment as an active entity is also overlooked in broader accounts of how societies construct memorials to mourn the dead, mark historical events, celebrate national origins, and commemorate war.[9] For geographer Kenneth Foote, memorials create a landscape "that is more than a passive reflection of a nation's civil religion and symbolic totems. Landscape is the expressive medium, a forum for debate, within which these social values can be discussed actively and realized symbolically."[10] While Foote accords the social dimension of memorials fluidity and change, their physical location within and interaction with active and historically contingent environments is somewhat underdeveloped.[11] In a sense, this approach arguably reproduces the intention of many memorials to present themselves (and the values they symbolize) as natural. As James E. Young argues, memorials "suggest themselves as indigenous, even geological outcroppings in a national landscape; in time, such idealized memory grows as natural to the eye as the landscape in which it stands."[12] This intention is true of a resistance monument in Gresse-en-Vercors, which self-consciously mirrors the imposing shape of Mont-Aiguille (figure 8.1).[13] I do not wish to suggest that scholars, such as Young, ignore the landscape in which the memorial stands, as this is certainly not the case.[14] I propose, however, that bringing an environmental perspective to memorials may help to dislodge more comprehensively a memorial's claim to naturalness.

An environmental history of memorials is a valid enterprise if we accept that memorials actively engage with their environment and in turn the environment actively engages with them. Nature, in its many forms, is appropriated to enhance the memorial experience. For instance, the Gilioli memorial above Vassieux-en-Vercors gains much of its impact from its mountainside location, commanding stunning views toward the *hauts plateaux* of the Vercors. Furthermore, contemporary memorial designs engage the visitor through sensual, as opposed to purely visual, experiences; some Holocaust memorials in particular create dark and dislocated spaces designed to unsettle the visitor.[15] Memorials thus create their own microenvironments while establishing their relation to a wider landscape.

Fig. 8.1 Resistance memorial at Gresse-en-Vercors. © Chris Pearson.

Moreover, memorials don't exist in an environmental vacuum. They are not immune from the effects of weather and encroaching vegetation, and they require active human intervention if they are to be preserved in a recognizable form. Sarah Farmer observes this happening in the ruined village of Oradour-sur-Glane, where weather erodes the ruins, softening and romanticizing them. They now "convey a mood of not unpleasant melancholy, rather than revulsion at horror." For Oradour's preservers, nature poses a "whole set of dilemmas—technical and aesthetic, emotional and political."[16] Elsewhere, as Andrew Charlesworth and Michael Addis bring to light, the preservers of memory at sites of former concentration camps are forced to grapple with the forces of nature and develop strategies of ecological management in their attempt to attain historical authenticity. Here too, nature acts in surprising ways; the increasing mole population leads to "molehills of grey ash and white bone" disrupting the newly created lawns at Auschwitz-Birkenau.[17] In light of such observations, I treat the relationship between memorial sites and the environment in the Vercors as reciprocal, active, and, at times, highly ambiguous.

The starting point for my environmental analysis of memorialization is the Vercors' "ravaged" and "martyred" postwar landscape (in the words of a former maquisard).[18] I suggest that the Vercors' memorials were not constructed within a blank landscape; rather, they are part of the process of rebuilding life in the Vercors and coming to terms with the events of 1944. I then turn to social constructions of the Vercors as a "natural fortress" that is now called upon to enclose and safeguard memory, before examining how the memorials appropriate the majestic landscape of the Vercors in order to glorify resistance and recreate the maquisard's experience of nature. Nature's agency, present throughout the memorialization process, is increasingly apparent in the final section of this chapter as the environment challenges the sites of resistance and their meanings. Repeated attempts to "ground" the maquisards' memory in the landscape of the Vercors are evident throughout. I use the term *ground* to suggest that memories of war have been invested in the rocks, soil, and memorials of the Vercors.

The War-Torn Landscape

Up until 1944, life in the Vercors had not been unduly disturbed by the war and the Vichy regime.[19] Indeed, after French military defeat in 1940, the Vercors seemed a world apart from the hardships and dangers of the occupation to the numerous refugees and maquisards who sought shelter there.[20] However, the German assault of 1944 shattered these illusions, transforming the Vercors both imaginatively and physically. For contemporary observers it seemed that the conflict had disrupted the normally productive relationship between the inhabitants and their environment. A group of Swiss journalists noted that "in the villages, in the pastures, on the irregular and sheep-filled higher plateaus, all is burnt, devastated, destroyed."[21]

It appeared that the German troops had stripped the Vercors bare. All that remained were "pastures without cows and sheep, bedrooms without floors, forests without echoes. If they could have done, [the soldiers] would have taken light and water with them." There was an uneasy ambiance in this pillaged land; "the atmosphere ... is of the most atrocious malignancy: avowed torture [was] perpetrated in the fresh air, in the middle of the day at one thousand metres of altitude, where, in normal times, vets would come to care for the livestock."[22] It was almost as if the violence had physically stained the mountains of the Vercors; the bodies of the dead "traced a bloody border around the summits."[23] This atmosphere did not go unnoticed by the authorities. The Prefect of the Isère reported that the military "engagements" had left a "widespread uneasiness and even a real terror" across the region.[24]

The scale and barbarity of the atrocities seemed unnatural. A publication produced by the Pionniers du Vercors, the most influential Vercors veterans association, emphasized the German troops' "bestiality," which was "without precedent."[25] The aforementioned Swiss journalists felt that the scale and brutality of the atrocities went beyond the wildness normally found in nature ("beasts don't take pleasure in the suffering of their victims"). They believed that the violence had "broken the limit of natural things to where good and evil merge together ... in front of the survivors, one understands the wickedness of nature and men."[26]

Contemporary reports also laid bare the physical destruction wrought by the war, outlining how German forces requisitioned livestock to prevent the local population from feeding maquis groups, adding to the already severe shortage of food supplies. As if to worsen the situation further, work in the fields had been put on hold due to a shortage of manpower.[27] Geographer Peter Nash carried out a study of this devastation, reporting that nearly all of Saint-Nizier-du-Moucherotte's chickens had been killed (the mayor claimed that the German troops had killed them "for the sake of it") and that rabbits and poultry had been taken from La-Chapelle-en-Vercors, while Vassieux-en-Vercors had lost nearly all of its large animals. In all, Nash estimated that the German army had requisitioned 25 percent of all livestock, 66 percent of horses, and 33 percent of pigs. Farmers had also endured the loss of substantial quantities of potatoes, wheat, and farm machinery.[28]

A particular challenge to agricultural life in the Vercors concerned the "strong, robust, and rustic" Villarde breed of cow, celebrated for its ability to both work in the fields and produce milk and meat. The war wiped out a quarter of the Villardes and (along with postwar agricultural modernization) almost led to the disappearance of the breed, which only recovered due to the introduction of conservation programs in the 1970s and 1990s.[29] The local population immediately felt the effects of the Villarde's demise, complaining that cows donated by Switzerland in the tragedy's aftermath were unsuited to life in the Vercors.[30] In contrast to the agricultural situation, what little we know of the fate of the Vercors' undomesticated animals remains on the level of speculation. In 1942, Marcel Couturier recorded that the brown bear was last spotted in 1937, 3 kilometers to the northwest of Saint-Martin-en-Vercors. He concluded that "thanks to its longevity, the bear is still part of the fauna of the

Creating the Natural Fortress 155

French Alps."[31] Further research, however, is needed to study the effect of war on the brown bear and the region's other fauna.

More is known about the built environment, which suffered heavily, even if the devastation was uneven across the massif and, as a whole, the Isère and Drôme were not as physically damaged by the war as *départements* in northern and eastern France.[32] Nevertheless, 573 buildings had been completely destroyed across the Vercors. In Vassieux-en-Vercors, only seven structures were left standing, with 97 percent of buildings razed. The figure was 95 percent in La-Chapelle-en-Vercors and 73 percent in Malleval, and in Pont-en-Royans bombing raids had "weakened the majority of houses."[33] The task of reconstruction in the Vercors was not inconsiderable, and the president of the Comité d'aide et de reconstruction de Vercors used the high level of localized destruction as a lever in negotiations to obtain supplies of wood, a material that was in high demand in post-Liberation France (it was essential to both the war effort and rebuilding the country).[34] Yet despite the expense, urgency, and difficulties involved in reconstruction, the memorialization of the maquis began almost immediately: the majority of memorials in the Vercors were erected between 1945 and 1950.[35] The Pionniers du Vercors spearheaded their construction, and by 1947 their efforts had already resulted in the inauguration of the cemetery at Saint-Nizier.[36] This suggests that remembering the dead was considered to be a key component of rebuilding life in the Vercors and restoring a united sense of purpose between the local population and the massif. Having resisted and suffered together, humans and nature combined, through the memorials, to perpetuate and strengthen memory.

Fortifying Memory

The image of the Vercors as a "natural fortress" has proved a recurrent and attractive one throughout the memorialization process, as diverse memories of the war are unified and enclosed within the massif's limestone cliffs. After the war, the Pionniers du Vercors proudly and publicly evoked the fortresslike qualities of Vercors with its "almost impenetrably high forests, its deep rock and cliff faces." It "lends itself admirably to the role of protector."[37] The PNRV, introducing itself in the pages of *Le Pionnier du Vercors* in 1974, described the area as "an impervious fortress," language that was sure to appeal to former maquisards.[38] Twenty years later, the rationale for the SNHRV also drew on the fortress image:

> The borders of the Vercors! Terribly abrupt and dominant. They impose their boldness and their rectitude over the surrounding valleys. The cliff faces drop straight down to the valleys. They attract the attention, they magnetize, they bewitch. It is not necessary to search any further for the reasons for resistance. It's there, inevitably, that

one day or another, the beautiful and cruel tragedy of a rebellious people would unfold. It's there, obviously, that one of the most poetic and symbolic pages of French history would imprint itself onto the limestone and spruce trees. This territory is a natural donjon.[39]

Such logic relies on a hefty dose of environmental determinism; the Vercors' topography is presented as the main reason for resistance. The mountains and their inhabitants naturally led to resistance.

Such ideas are rooted in the Plan Montagnards, a strategic plan developed in 1942 by keen alpinist Pierre Dalloz, who conferred a military purpose on the Vercors. Dalloz believed that "it is a fact that this massif, with its rare and difficult access routes, its immense forests, its remote pastures, its sheepfolds, its little known network of caves, could provide the most secure refuge for outlaws." The massif's resistance potential came "as much from the configuration of its soil as the independent and courageous character of its sons." The Vercors radiated resistance spirit. Indeed, it was this quasi-mystical power that Dalloz claimed inspired the Plan Montagnards in the first place: "I opened my door. I breathed in the night's cold air. The familiar almond tree, stirred by the wind, brushed the starry sky with its foliage. . . . The mountains were straight ahead. There was the Vercors. I reflected for a long time. The darkness was an accomplice. The moment was heavy for me, full of responsibility, resolution, and hope."[40] The mountains were waiting, encouraging Dalloz to realize their potential. He proposed that this "fortress" could shelter a substantial number of maquisards, as well as offering Allied forces a secure aerial base within France for the anticipated landings in Provence.

The maquisards' difficult ascent of the Vercors strengthened the idea of a natural fortress that the Plan Montagnards represented. One of them recalled the journey: "the road climbed steeply alongside a small, calm stream, gently bubbling over the rocks. This must have been a torrent in the springtime. Our chocolate and bread had long gone, and we were thirsty. We carried on across wild gorges, without coming across a house or a person."[41] And after the war, the fortress image serves to naturalize resistance; the maquis were merely heeding nature's call to come and fight. In this view, the massif itself is enough to explain and remember the resistance. Indeed, Dalloz, opposing plans to build a new memorial in the 1970s, urged that the massif be left alone. For him "the Vercors is itself a monument, the most grandiose of monuments."[42]

With hindsight, however, the reliance on the "natural fortress" can at best be described as a noble failure. Leaving aside the supposed Allied and Free French "betrayal" of the Vercors, a controversy heightened by Cold War political posturing, the leaders of the maquis had considerably overestimated the area's defensive strength.[43] Historian Paul Silvestre argues that the "natural fortress" generated a false sense of security that obscured the difficulties involved in living on the massif, such as the paucity of the water supply. He believes that maquis groups best used mountainous areas as a base for sabotage attacks on enemy forces. Above all, the units had to be mobile and dispersed, a view backed up by contemporary Allied intelligence.[44] In a similar vein, resistance historian Henri Noguères outlines the Vercors' defensive frailties. Indeed, by mid-June, German troops had already succeeded in opening up an access route through Saint-Nizier. The central plateau was also accessible by road from the southeast, and even the apparently impenetrable eastern ridge contained a series of mountain passes that Axis alpine troops were trained to breach. Noguères also points to strategic errors on the part of the maquis leaders, such as failing to adequately defend Vassieux-en-Vercors and underestimating the enemy's strength.[45]

The persistence of the "natural fortress" myth can be explained from a variety of angles. For a start, it helps to bind memories of the maquis into a unified whole, thereby subsuming the diversity of experience and the tensions that existed between various maquisards. The Pionniers du Vercors successfully transcended the diversity of the Vercors in their desire to speak "of the Vercors, in the name of the Vercors." As Philippe Barrière observes, their logo of a chamois astride the word "Vercors" is a recurrent motif of the memorials and "practically functions as a label, a sign of quality, encompassing and equalizing all the experiences (combats, martyrs, French, foreign, military, civilian) at the heart of the largest 'vercorienne' memory possible."[46] More recently, the SNHRV has sought to standardize the sites, proposing to plant a yew tree by the memorials in order to "create an identical symbol for all of the massif."[47]

Furthermore, the diversity of the region is surmounted by the natural fortress image. It was only in the early twentieth century that geographers marked out what we now consider to be the Vercors as a single, coherent space (but not without controversy within the discipline), collapsing topological, climatic, and administrative differences between north and south.[48] The shared (if uneven) experience of the war strengthened this tendency to treat the Vercors as a definably separate space. In the present day, the natural fortress image further extends this sense of separateness and now serves to protect the Vercors from the tides of history, giving the area (and its resistance history) a timeless and eternal quality. This has been a persistently attractive idea. In 1927, writer Albert Marchon rhapsodized about the Vercors' "empty skies, clear of all the past" and the "pure, clean lines" of its forests, which recalled the "temple of Arcadia."[49] The Vercors' environment, however, has never been divorced from history, being very much shaped by human activity in the form of agriculture and tourism.[50] Nor had it been a military-free zone before the war. In the 1930s, conservationists had opposed military maneuvers conducted in the region by the French army and called for the creation of a national park to protect the Vercors' central plateau, which was a "paradise" for botanists and geographers.[51]

Just as the environment of the Vercors has been shaped by human activity, so, too, does its meaning change in accordance with evolving human ideas about nature. For instance, the general rise of environmentalism in France has "greened" the "natural fortress."[52] During the war, the Vercors' main purpose was to shelter and protect the maquisards gathered between its walls. It is now called upon to safeguard their memory *and* protect the area's natural habitats and rural ways of life. As Anne Sgard argues, the image of the "natural fortress" increasingly rests on its protection of "preserved nature and mountain traditions."[53] The rhetoric and publicity of the PNRV taps into the discourse of environmental protection, proposing that within this "veritable natural citadel of limestone . . . wooded plateaus and valleys, shaped by agriculture, shelter a remarkable fauna and flora."[54] The potential of this "remarkable" nature has not been lost on the creators of memorials who expressly appropriate it to enhance the memorial experience.

The Glorious Experience of Nature

From the immediate postwar period onward, the Vercors' memorials have drawn on nature for dramatic effect and authority. For instance, the cemetery at Saint-Nizier, situated on the site of sustained fighting between June 13 and 15, 1944, offers the visitor "one of the most beautiful panoramas in the Alps," according to the official postcard. However, the layout of the cemetery draws the visitors' attention away from this view toward the imposing cliff faces of Le Moucherotte. These vertiginous rocks mirror the grayness of the cemetery while the mountain's green pines echo those of the cemetery. Consequently, it seems to me that the visitor is invited to compare resistance with the rugged mountain, giving it an aura of solidity and strength (figure 8.2).

The cemetery at Vassieux-en-Vercors (inaugurated in 1948) appropriates its natural surroundings in a similar way. The layout directs the gaze away from the tragedy that occurred in the village toward a mountainous backdrop, while the evergreen trees in front of the wooden crosses remind the visitor that life continues thanks to the sacrifice of the fallen. At both these sites the majestic landscape echoes the magnitude of the sacrifice of the dead and the causes for which they died, namely those of freedom and France.

A similar phenomenon is repeated across the Vercors. The Stations of the Cross, which trace a pilgrimage from Villard-de-Lans up to the ruined village of Valchevrière, appropriate their surroundings to emphasize a spiritual dimension to death. Inaugurated in 1948, these memorials represent the passion, sacrifice, and suffering of both Christ and the maquisards and were designed to "last for centuries and fix memory in an indestructible material." In doing so, they purposefully maximize their natural surroundings; "so that the work is beautiful, the most picturesque sites were chosen, and each oratory is adapted to the landscape."[55] Not all of these memorials are situated on sites where maquis fighters lost their lives, as aesthetic factors influenced the positioning of the monuments. As with the aforementioned cemeteries, the surrounding landscape assumes a central role in the grounding and fixing of the memory of the maquis.

Creating the Natural Fortress

Fig. 8.2 Cemetery at Saint-Nizier. © Chris Pearson.

The recourse to the natural splendor of the Vercors is not entirely divorced from the maquisards' wartime experience of nature (after all, the name *maquis* derives from the dense trees and scrubland in which they sheltered and fought). One maquisard remembers that "the weather was magnificent in this month of June. The sky is a resplendent blue. It feels deliciously good at 1,000 meters of altitude. And above all, up there, on the verdant summits, one feels free." Another recalls a "magnificent day." The men, for the most part bare-chested, "exposed their bodies, already tanned by the July sunshine."[56] A revitalized masculinity apparently blossomed in the mountains due to a heady combination of fresh air and liberty. The photos of Marcel Jansen, a young maquisard and budding photographer, visualize these feelings of youthful freedom and, were it not for the guns that the young men carry, would convey the impression of a camping or hiking holiday.[57] For some, *prendre le maquis* was about more than conflict and contained a spiritual element; the Vercors was "a landscape where you could shelter for three months of solitude and where you could discover something about yourself, and also about nature."[58]

Nature, in all its glory, created an intoxicating sense of freedom. One resister recalls that they "were . . . no longer part of the French State, *hein*, we were completely separate. That was something quite sensational."[59] Road signs declared, "*Ici commence le pays de la Liberté*" ("Here begins the land of freedom"), and a large tricolor flew from the summits. This freedom was unforgettable, according to the resistance journal *Aux Armes!*—"one breathed the air of freedom, one practiced a fraternity, an honor, which no one who was there will ever forget."[60] The Vercors became a landscape of liberty and resistance, with historian Anna Balzarro suggesting that the mountains symbolized freedom while fascism and oppression were left behind on the plains.[61] This symbolism continues to the present day. The creators of *The Alps at War* exhibition at Grenoble's Museum of Resistance and Deportation propose that Alpine solidarity and freedom during the war represented a different kind of Europe from that of Auschwitz.[62] In addition, words now used to describe the maquis itself—"solitude, strangeness, toughness, humility, quest, taste of freedom"—correspond remarkably with those associated with the resisters who took to the *garrigue*.[63]

However, hunger and hardship are also remembered among the sunshine and healthy outdoor activities. Although maquisards spent their days chopping wood, fetching water, searching for food, and singing in the evening, the routine could become tiresome. At times it also seemed that nature conspired against them, heightening their sense of danger and exhausting them with the exertions of mountain life. Gilbert Joseph remembers walking in the darkness: "the physical effort transformed the acoustics: all of nature seemed to buzz with a thousand stridulations of insects. But it was just the pulsations of our arteries and the accelerated beating of the heart which bored through to the bottom of our being."[64] The night was sinister, its silence "pierced by the cries of foxes and other beasts" who populated the forests. The intense stillness of the forest oppressed the maquisards as they hid from German troops.[65]

The 1994 memorial at Col de La Chau recreates for the visitor both the light and dark sides of this experience of nature and, more generally, the experience of resistance and defeat, although its final emphasis is very much on the Vercors' redemptive qualities. The dark, cold, and claustrophobic atmosphere of the exhibition halls leads the visitor out onto a bright balcony with spectacular views over the Vercors. This contrast between light and dark is a deliberate attempt to show the heroic/tragic elements of the Vercors. As the rationale for the memorial suggests, the "exit toward the light, a symbol of freedom, is a moment of confrontation between a landscape and its history. . . . the memorial reminds us of the two most contradictory faces of our humanity, darkness and light."[66] The choice of location is key to the memorial strategy. No major resistance activity took place on the actual site of the memorial; its position was selected for its view over the massif (figure 8.3).[67] This consideration, and the weight of the PNRV, overrode the opposition of Vassieux-en-Vercors' mayor, who was concerned about the difficulties of access up to the memorial during winter and the possibility that it would divert tourists away from the village (the memorial being recognized as a highly important resource for tourism and the local economy). Moreover, he felt, the scale of Vassieux-en-Vercors' losses in 1944 meant it should be "granted the principal infrastructure" of the SNHRV.[68]

Creating the Natural Fortress

8.3 The memorial at Col de la Chau, 1994. © Chris Pearson.

This memorial also attempts to ground memory in the most robust way possible. Its very form, the result of a 1993 architectural competition, aims to integrate itself into its surroundings as its shape follows the curve of the combe. In addition, the memorial is covered with local vegetation (junipers and pines) so that it blends in more fully with the site. In a sense, the memorial's design recognizes memory's fragility, requiring that its physical traces be reinforced. Indeed, the SNHRV justifies itself with the words of resister and writer Jean Bruller (nom de plume Vercors): "when memory weakens, when, like a fragile cliff, it begins to be eaten away by the sea and the weather, to collapse into the depths of oblivion, it is time to gather up what is left before it is too late."[69] It is here that we confront the internal contradiction at the heart of grounding memory, as nature itself, memory's supposedly eternal repository, contains the potential to radically transform, even destroy, the revered physical remains of 1944.

Nature's Challenge to Memory

Memorials do not exist in an environmental vacuum, nor can nature automatically be relied upon to bear witness to human events. Indeed, in the immediate aftermath of the tragedy, nature almost disguised the atrocities that were committed in the Vercors: "as if to delay the laying bare of so much horror, the luminous mountain landscape seemed peaceful around the ruins, in the surprising silence of the pastures without herds, in the immense calm of a land swelled with the dead, apparently left to its natural tranquillity."[70] Today, the relative quiet of Vercors (at least in comparison to nearby Grenoble) presents a jolting contrast to the violence that unfolded in the landscape.

Moreover, nature is oblivious to the human desire to preserve the past. In fact nature may be more inclined toward forgetfulness, as the site of the former internment camp of Saliers (near Arles), now reverted to an unmarked field, clearly shows. In the Vercors, environmental factors are slowly but surely transforming the historical remains of 1944. Weather erodes the metal shells of the SS gliders that landed at Vassieux-en-Vercors and are now on display in the village. Preservation efforts have not been entirely successful, and rust breaks through the layers of paint applied to the gliders. At Valchevrière, nature is in the process of reclaiming the houses that were left in a ruined state after German troops had laid waste to them. Vegetation is now poised to swamp the ruins (figure 8.4).

Steps are taken to mitigate nature's onward march. The environmental services of the commune of Villard-de-Lans manage the ruins, clearing trees and other vegetation if necessary and replanting some species, such as walnut and lime trees. A desire to maintain the "natural aspect" of the site informs this management.[71] Ironically, active human intervention is needed to preserve the "natural state" of the ruins, because if nature were left to run riot, the site would change beyond all recognition. This paradox threatens to seriously derail the whole strategy of grounding memory, as nature, the supposed preserver of memory, becomes the destroyer of memory.

The Rester-Résister garden of memory in Vassieux-en-Vercors (1994) further demonstrates the ambiguity inherent in mobilizing nature to proliferate and preserve memory. In this memorial space rectangular panes of glass shoot

Creating the Natural Fortress 163

Fig. 8.4 *In among the weeds: the ruins of Valchevrière.* © *Chris Pearson.*

vertically up from the ground, their number corresponding to that of civilian deaths during the massacre at Vassieux-en-Vercors. Their clean, straight lines contrast with the row of older graves to the rear of the site, which are weathered and covered in lichen. As with many of Vercors' other memorials, nature is a sphere through which human loss is remembered and understood, as the purity of the martyrs' sacrifice is reflected in the landscape visible through the glass. A plaque outside the garden explains the panes' significance:

> Pane of glass: symbol of the fragility of life, so quickly shattered, so quickly destroyed . . .
> Pane of glass: symbol of the purity of the soul of the martyrs and their ideal.
> Pane of glass: symbol of transparency. . . . through which opens the image of the plateau where the stones and the wind recount—if one knows how to listen to them—a terrible and glorious history.

The winds and stones are asked to share the burden of memory, but by themselves they reveal nothing to the onlooker about the history of the maquis. The memorials are needed to frame and contextualize this landscape. However, this frame is liable to crack. On a visit to the site in July 2004 the garden was shut to visitors, as the mayor was concerned that the frequent breaking of panes constituted a safety hazard. Theories put forward for the breakages include strong winds, wayward birds, or vandals. Whatever their cause, the breakages show how environmental and social conditions can undermine the will to remember.

The celebrated natural environment also attracts visitors to the area who may be more interested in leisure activities than resistance history. As noted above, tourism has long since existed in the Vercors, although its scale has intensified in recent years in line with its increasing importance to the local economy, especially in the more economically disadvantaged southern region.[72] This development has left its mark on the landscape. In the first decade of the PNRV, fifty hotels and twenty ski resorts were constructed.[73] A cross-country ski kiosk now sits on the route of the Stations of the Cross, and husky dog rides are available in winter 200 meters above the memorial at Col de la Chau. As these cases demonstrate, meeting the demands of modern tourism does not always go hand in hand with reverence for the maquis heritage.

At times, the differing priorities of the tourists and former maquisards come head to head. Publicity material now sells the Luire cave (where German troops murderously disbanded a maquis field hospital) on the basis of its resistance history, its spectacular rock formations, and its "real and savage beauty." The site's speleological and resistance elements sit uneasily together according to veteran Paul Jansen, who expressed his concern that the demands of tourism, such as improving access to the cave and providing parking facilities, threatened to transform this "sacred" place. He believes that "the comparison between the rare photos taken at the time and those of today show that the changes to the cave have removed the character which history conferred onto it." He reports that some tourists (presumably hoping for a taste of authentic resistance history) are disappointed by the state of the cave and feel "tricked."[74] Jansen's views are not without reason. The cave's sacredness and solemnity are

Creating the Natural Fortress

somewhat undermined by the kiosk toward the rear of the cave selling guided tours and postcards.

Elsewhere, a former maquisard bitterly recounts how he came across the following message left by a tourist in a refuge near the Pas de l'Aiguelle memorial: "very pleased to come here to admire this magnificent landscape. Shame it's spoilt by this monument." Sardonically berating this visitor's lack of respect for the eight bodies that lie at the site and noting how the memorial is purposely placed there so as to be clearly visible from all directions, the veteran makes the following point: "we are not just anywhere here. This ground has become historical, for having known both glory and suffering."[75] Reconciling this landscape's historical and natural uses is becoming increasingly fraught with difficulties, as those who have sought to ground the memory of the maquis are discovering how unpredictable that very ground (and its meanings) can be.

Conclusion

This environmental history of memorialization has outlined how memories of 1944 have been invested in the rocks, soil, and cliff faces of the Vercors, from the immediate postwar period onward. The memorials, enclosed within the celebrated "natural fortress," frame their surrounding environment for the visitor, and in doing so, naturalize, glorify, and legitimize resistance activity. In the Vercors, the grounding of memory is an attempt to escape the transitory nature of human existence and the fragility of remembrance. As a consequence, the land has been fashioned into a reliable witness, while succor and redemption are sought in the Vercors' seemingly stable and eternal bulk. The nature celebrated here is not so much that of la douce France (a thoroughly cultivated land domesticated "with understanding gentleness rather than massive attack"[76]), but a wilder landscape where forestry and mountains represent that which is supposedly pure and true. Yet this memorial landscape is far from natural. It is culturally produced, principally by the Pionniers du Vercors, and memory's current physical representations are the result of numerous human decisions about whom to remember, and where, how, and why to memorialize this memory.

As the past recedes and the maquis generation passes on, the drive to reinforce war's physical presence in the contemporary landscape has intensified. This tendency culminates in the memorial at Col de la Chau, which aims to recreate the maquis' redemptive experience of nature, and represents the boldest attempt yet to ground memory. Nature, however, is proving to be a far from reliable witness. Physically deteriorating sites and the increasing demands of contemporary tourism now represent the greatest challenges to the sanctity and authenticity of the Vercors' resistance heritage. More difficult decisions, therefore, await the preservers of memory as they try to reconcile the grounding of memory with the unpredictable reality of nature, while combining the need to commemorate the dead and safeguard the values of the resistance.

The Vercors has been shaped into a symbol of freedom, a foundation of contemporary French republicanism. As Philippe Mestre, the former Minister of Veterans and Victims of War implored his fellow citizens in 1994:

> Come to the Vercors.
> Understand the close links which unite this exceptional territory and its fighters.
> Understand what we owe to those who, up there, gave everything.
> May their memory help the French people to recognize the founding values of the republic.
> May the memory of the Vercors inspire goodwill, of which peace has so much need.[77]

The worry must be that in fifty to a hundred years time the physical forms of memory will be overgrown with weeds and the Vercors' landscape and its meanings irredeemably transformed by environmentalism and tourism. Furthermore, the amount of effort, time, and money accorded to conserving this resistance heritage may become increasingly uncertain as it jostles for attention among the calls for environmental protection and leisure-based tourism. The intricate relationship between the Vercors' memorials and their environment looks set to become progressively more ambiguous.

Notes

1. Quoted in *Le Dauphiné Libéré*, July 22, 2004, 2.

2. Quoted in Joseph La Picirella, *Témoignages sur le Vercors Drôme et Isère* (Lyon: Rivet, 1969), 383.

3. The number three hundred comes from Olivier Vallade, *Des combats au souvenir: Lieux de résistance et de mémoire—Isère et Vercors* (Grenoble: Presses universitaires de Grenoble, 1997), 5.

4. The PNRV's other roles are maintaining the local economy and "developing harmony between man and the environment." See "Un territoire d'exception," at http://www.pnr-vercors.fr/parc/index.html, accessed August 2, 2004.

5. The most comprehensive account is Gilles Vergnon, *Le Vercors: Histoire et mémoire d'un maquis* (Paris: Atelier, 2002). See also Anna Balzarro, *Le Vercors et la zone libre de l'alto Tortonese: Récits, mémoire, histoire* (Paris: Harmattan, 2002); Pierre Bolle, ed., *Grenoble et le Vercors: De la résistance à la libération* (Lyon: Manufacture, 1985); Paul Dreyfus, *Vercors: Citadelle de liberté* (Grenoble: Arthaud, 1969); Patrice Escolan and Lucien Ratel, *Guide-Mémorial de Vercors resistant: Drôme-Isère, 1940–44* (Paris: Cherche Midi, 1994); Paul Silvestre, "STO, maquis et guérilla dans l'Isère," *Revue d'histoire de la deuxième guerre mondiale* 130 (1983): 1–50.

6. Michael Pearson, *Tears of Glory: The Betrayal of Vercors, 1944* (London: Macmillan, 1978), ix.

7. For the "Great War's" commemorative landscapes, see Michael Heffernan, "Forever

England: The Western Front and the Politics of Remembrance in Britain," *Ecumene* 2, no. 3 (1995): 293–324; David W. Lloyd, *Battlefield Tourism: Pilgrimage and the Commemoration of the Great War in Britain, Australia, and Canada, 1919–1939* (Oxford: Berg, 1998); Mandy S. Morris, "Gardens 'For Ever England': Identity and the First World War British Cemeteries on the Western Front," *Ecumene* 4, no. 4 (1997): 410–34; and George L. Mosse, *Fallen Soldiers: Reshaping the Memory of the World Wars* (New York: Oxford University Press, 1990), 107–25.

8. For instance, Henry Rousso limits his classic study of the occupation's memory to films, political scandals, and legal trials; see *The Vichy Syndrome: History and Memory in France since 1944*, trans. Arthur Goldhammer (Cambridge, Mass.: Harvard University Press, 1991). See also Omer Bartov, "Trauma and Absence: France and Germany, 1914–45," in *Time to Kill: The Soldier's Experience of War in the West, 1939–1945*, ed. Paul Addison and Angus Calder (London: Pimlico, 1997), in which Bartov argues that the occupation left few notable physical traces. The exceptions relevant to World War II's physical aftermath include Serge Barcellini and Annette Wieviorka, *Passant, souviens-toi! Les lieux du souvenir de la Seconde Guerre Mondiale en France* (Paris: Plon, 1995); Sarah Farmer, *Martyred Village: Commemorating the 1944 Massacre at Oradour-sur-Glane* (Berkeley: University of California Press, 1999); and Danièle Voldman, *La reconstruction des villes françaises de 1940 à 1954: Histoire d'une politique* (Paris: L'Harmattan, 1997).

9. See Alan Borg, *War Memorials: From Antiquity to the Present Day* (London: Leo Cooper, 1991); Nuala Johnson, "Cast in Stone: Monuments, Geography, and Nationalism," *Environment and Planning D: Society and Space* 13 (1995): 51–65; Edward Tabor Linenthal, *Sacred Ground: Americans and Their Battlefields* (Urbana: University of Illinois Press, 1993); and Jay Winter, *Sites of Memory, Sites of Mourning: The Great War in European Cultural History* (1995; rpt. Cambridge: Cambridge University Press, 2003).

10. Kenneth E. Foote, *Shadowed Ground: America's Landscapes of Violence and Tragedy* (1997; rpt. Austin: University of Texas Press, 2003), 292.

11. For the view of nature as disruptive and unpredictable, see Donald Worster's "Nature and the Disorder of History," in *Reinventing Nature? Responses to Postmodern Deconstruction*, ed. Michael E. Soulé and Gary Lease (Washington, DC: Island Press, 1995), 65–85.

12. James E. Young, *The Texture of Memory: Holocaust Memorials and Meaning* (New Haven, CT: Yale University Press, 1993), 2. In contrast, certain memorials, such as Daniel Libeskind's Jewish Museum in Berlin, actively jar with their surroundings, generating their "own sense of a disquieting return, the sudden revelation of a previously buried past" (James E. Young, *At Memory's Edge: After-Images of the Holocaust in Contemporary Art and Architecture* [New Haven, CT: Yale University Press, 2001], 154).

13. Philippe Barrière, "'Au nom de la mémoire . . .': Les associations grenobloises d'anciens combatants et victimes de guerre à la libération (1944–1947)," *Guerres mondiales et conflits contemporains* 205 (2002): 34–53, esp. 53.

14. See, for instance, Young, *Texture of Memory*, 7–8.

15. James Winchell, "Holocaust Memorials in France: A Walking Tour for the Body-at-Risk," *Contemporary French Civilization* 20, no. 2 (1996): 304–340, esp. 305, 335; Young, *Texture of Memory*, 257–58, 342.

16. Sarah Farmer, "Oradour-sur-Glane: Memory in a Preserved Landscape," *French Historical Studies* 19, no. 1 (1995): 42–43.

17. Andrew Charlesworth and Michael Addis, "Memorialisation and the Ecological Landscapes of Holocaust Sites: The Cases of Auschwitz and Plaszow," *Landscape Research* 27, no. 3 (2002): 229–51.

18. Paul Jansen, "Récits, témoignages, histoire . . . la reconstruction du Vercors," *Le Pionnier du Vercors*, n.s. (September 1991): 12.

19. Vergnon, *Le Vercors: Histoire et mémoire d'un maquis*, 31.

20. See François Boulet, "Montagne et résistance en 1943," in *Mémoire et histoire: La resistance*, ed. Jean-Marie Guillon and Pierre Laborie (Paris: Éditions Privat, 1995), 261–69.

21. Albert Béguin et al., *Le Livre noir du Vercors* (Neuchâtel: Ides et calendes, 1944), 22.

22. Pierre Courthion, "L'atmosphère," in Béguin et al., *Livre noir*, 23.

23. Archives Départementales de l'Isère (hereafter, ADI) 13R 1043, *Aux armes!* July 10, 1945, 3.

24. ADI 13R 990, letter from the prefect of the Isère to the regional prefect in Lyon, July 29, 1944.

25. ADI 4° 330, program for the performance of the opera *Mirreille*, produced by the Amicale des Pionniers et des Combattants volontaires du Vercors, June 20, 1948.

26. Albert Béguin, "Au seuil de l'Enfer" in Béguin et al., *Livre noir*, 51; Courthion, "Atmosphère," 25.

27. ADI 13R 990, M. Duboin, "Rapport sur les événements survenus dans la region de Villard-de-Lans," July 26, 1944; M. Duboin, "Rapport sur la situation dans les communes de St Nizier, Villard-de-Lans, Lans, Corrençon, Meaudre, et Autrans," August 1, 1944.

28. Peter H. Nash, "Le massif du Vercors en 1945: Étude sur les dévastations causes par l'armée allemande dans une région alpine de la France et de leur effets sur les traits géographiques," *Revue de Géographie alpine* 34, no. 1 (1946): 91–93.

29. Denis Chevallier, *Le temps des Villardes: Une race bovine de montagne* (Lyon: La Manufacture, 1986), 53; "100 ans: C'est le bel age de la Société d'élevage bovin de Villard-de-Lans," *Le Dauphiné Libéré*, August 31, 1975, back page; and "La Villarde: Le renouveau d'une race oubliée," *Le journal du Parc* 39 (Winter 2003): 6.

30. Nash, "Massif du Vercors," 92.

31. Marcel A. J. Couturier, "L'ours brun dans les Alpes français: Sa chronologie actuelle. Le dernier ours tué (Maurienne, Savoie), le dernier ours vu (Vercors, Drôme), *Revue de Géographie alpine* 34 (1942): 788–90.

32. ADI 16R/1, "Montant des dommages de guerre en valuer 1949–1950," map no. 50, produced by the Ministére de la reconstruction et de l'urbanisme.

33. Nash, "Massif du Vercors," 90; ADI 13R 990, letter from the prefect of the Isère to the regional prefect in Lyon.

34. ADI 6P1/22, letter from president of the Comité d'aide et de reconstruction de Vercors to prefect, December 7, 1944; letter from Director General of Eaux et Forêts to Regional Conservateurs, November 14, 1944.

35. Vallade, *Combats au souvenir*, 14.

36. Barrière, "Nom de la mémoire," 47–48.

37. ADI 4 330, "Programme for Mirreille."

38. *Le Pionnier du Vercors*, 30th anniversary edition, 1974.

39. "Un site predestiné," *Site national historique de la résistance en Vercors*, brochure produced by the PNRV (n.d.), 17.

40. ADI 13R 1043, "Naissance des maquis du Vercors par Pierre Dalloz," 2–4, 7.

Creating the Natural Fortress

41. Jean Dacier, *Ceux du maquis, coups de main et combats: L'épopée d'une compagnie d'F.F.I de Vercors* (Grenoble: B. Arthaud, 1945), 60.
42. Quoted in Vergnon, *Le Vercors: Histoire et mémoire d'un maquis*, 187.
43. For more on this controversy, see ibid., 13–14, 157–77.
44. Silvestre, "STO, maquis et guérilla," 11, 47; ADI 13R 1043, "Note sur les Alpes de Provence," rédigé à Londrès—remis au Commandant Manuel, chef du B.C.R.A.L., May 12, 1944, 1.
45. Henri Noguères, *Histoire de la Résistance de 1940 à 1945*, Tome 5, *Juin 1944–Mai 1945* (Paris: R. Laffont, 1981), 342–43, 375.
46. Barrière, "Nom de la mémoire," 47.
47. "Les lieux de mémoire," *Site national historique*, 30.
48. See J. Offner, "Les étages de végétation du massif du Vercors," *Revue de Géographie Alpine*, 7, no. 1, 1920, 125; and Anne Sgard, "L'invention d'un territoire," *Un hors-série de l'Alpe: Vercors en questions* (Meylan, 2001), 45–46. Traditionally, the Vercors was considered to be a tightly defined area around Vassieux-en-Vercors.
49. Albert Marchon, *Le Vercors* (Paris: Émile-Paul frères, 1927), 100. Historian Philippe Hanus challenges this view of history, arguing that the "middle mountain has never been out of, nor even on the margins, of history" ("Briser le mythe de la citadelle," *Hors-série de l'Alpe*, 17).
50. See André Micoud, "Un monde sans hommes?" and Michel Wullschleger, "La saga du Vercors," *Hors-série de l'Alpe*, 8–11, 18–32.
51. G. Bossière, "Un Parc National dans le Vercors," *La Nature* 3015, December 15, 1937, 584.
52. For more on the emergence of French environmentalism, see Michael Bess, *The Light-Green Society: Ecology and Technological Modernity in France, 1960–2000* (Chicago: University of Chicago Press, 2003).
53. Sgard, "Invention d'un territoire," 50.
54. *Terre d'accueil: Le Parc Naturel Régional du Vercors, vif de nature, vivant de culture*, PNRV leaflet, 2001.
55. M. Le Chanoine Jacques Douillet, *Valchevrière: Le chemin de croix du Vercors* (Lyon: Lescuyer [196?]), 3.
56. Quoted in Balzarro, *Vercors*, 180–81.
57. Marcel Jansen, *Reporter au maquis: Les photos de Marcel Jansen* (Valance: Peuple Libre, 1994), 35–63.
58. Alain le Ray, quoted in Eric Aeschimann, "Héros du plateau," *Liberation*, May 17, 2004, at www.liberation.fr/page.php?Article=206301, 2.
59. Interview with Paul Borel, *Enquêtes sur la mémoire orale des anciens du Maquis du Vercors: 1er rapport intermédiare* (Avignon, 1992). The Vichy regime called itself the "État français" to distinguish itself from the "République français."
60. *Aux armes!*, 3.
61. Balzarro, *Vercors*, 180–81.
62. See the exhibition catalogue, Gil Emprin and Jacques Loiseau, *Alpes en guerre, 1939–1945: Une mémoire en partage* (Veurey/Grenoble: Musée de la Résistance et de la déportation de l'Isère, 2003), 3.
63. Conseil Général de la Drôme, *Mille tendresses: Grands espaces naturels de la Drôme* (Valance, 2000), 36.

64. Gilbert Joseph, *Combattant du Vercors* (Paris, 1972), 12, 21.

65. Brun-Bellot, "Ambel, premier maquis du Vercors," in *Le Vercors raconté par ceux qui l'ont vécu*, ed. Association nationale des Pionniers et combatants volontaires du Vercors (Grenoble, 1990), 80; Dacier, *Ceux de maquis*, 135.

66. *Le memorial de la Résistance en Vercors*, PNRV leaflet, undated.

67. "Le Mémorial du col de la Chau," *Site national historique*, 24.

68. Extrait du registre des délibérations du conseil municpal, Commune de Vassieux-en-Vercors, August 3, 1992, 1–2.

69. Quoted in *Le site national historique de la Résistance en Vercors: Les chemins de la Liberté*, PNRV, leaflet, undated.

70. Béguin, "Au seuil de l'enfer," 38.

71. Personal communication with the Maison du patrimoine, Villard-de-Lans, August 17, 2004.

72. Louis Reboud and Michel Wullschleger, "Les leçons d'un recensement," *Hors-série de l'Alpe*, 12–15.

73. Marc Ambroisé, "Rendu heurs et malheurs des parcs naturels: Le Vercors tenté de concilier archéologie, tourisme, et mise en valeur économique," *Le Monde*, July 16, 1980, 25.

74. Paul Jansen, "Un lieu sacré: La grotte de la Luire," *Le Pionnier du Vercors*, n.s. 76 (September 1991): 1.

75. "Le mot du Chamois," *Le Pionnier du Vercors*, n.s. 2 (April 1973): 14–15.

76. Armand Frémont, "The Land," in *Realms of Memory: Rethinking the French Past*, vol. 2, *Traditions*, ed. Pierre Nora, trans. Arthur Goldhammer (New York: Columbia University Press, 1997), 25.

77. Quoted in *Site national historique*, 2.

CHAPTER 9

WARTIME DESTRUCTION AND THE POSTWAR CITYSCAPE

Jeffry M. Diefendorf

For more than a decade, environmental historians have asserted the need to link their discipline with the history of urban planning. In the modern era, they contend, the shape of the built environment is intrinsically related to changes in the biosphere.[1] Various forms of pollution, the exploitation of organic resources, epidemics, and natural catastrophes should be examined not only as ecological phenomena but also as determinants of urban form. Town planners and architects, drawing on the vocabulary developed by geographers to describe natural and man-made environments, refer to *cityscapes* and *urban landscapes* as counterparts to the conventional notion of landscape as a space unmarked by human activity. Urban historians, for their part, have considered the impact of natural disasters on the built environment and, in separate studies, how cities have been changed by wartime mobilization and destruction.[2] What is missing here is a connection between cities, wars, and the natural environment.

The purpose of this chapter is to demonstrate this connection by exploring the ways in which urban planners, particularly in Germany but elsewhere as well, responded to the challenges posed by the two world wars. Obviously the bombing of cities had a dramatic impact on the built environment, but the main focus here will not be destruction so much as the extraordinary opportunities that war seemed to provide for planners to reshape cities in ways that could better harmonize the built and natural environments.

If one looks at the period from around 1890 up to around 1960, urban-planning models can be grouped within three clusters: variants on the garden city, variants on the city beautiful and monumentalism, and variants on modernist functionalism. The first cluster includes true garden cities, as proposed by Ebenezer Howard and constructed at Letchworth in Britain, and garden suburbs, such as Hellerau outside of Dresden. The building of new towns and town extensions had no measurable impact on the cores of existing cities. The second cluster includes proposals and actual attempts to transform cities by constructing large buildings and creating broad avenues. New civic centers, the completion of the buildings on Vienna's Ringstraße, as well as the ideologically driven redesign plans of Mussolini, Hitler, and Stalin fit here. The modernist, functionalist cluster of models includes not only the Charter of Athens produced by Le Corbusier and the members of the Congrès Internationaux d'Architecture Moderne (International Congress of Modern Architecture, known as CIAM) but also most of the wartime and early postwar reconstruction plans. For that reason these latter concepts of urban design are of most interest here.

All three clusters of planning models coexisted after World War I, which means that there were significant continuities in ideas propagated by individuals and institutions over the next four decades, continuities that persisted in spite of sharp ideological differences embodied by extreme political regimes and the enormity of the urban destruction wrought during World War II. All three clusters shared certain premises. All, whether on the political left or right, regarded existing cities as unhealthy in the long term and dangerous in the short term. Born in the age of rapid population growth and industrialization, such cities were believed to promote life-threatening physical disease (medical epidemics like cholera and tuberculosis) and civilization-threatening social ills (criminality, amoral behavior, and political radicalism.) The cities of stone, to use Werner Hegemann's phrase, were overcrowded, without enough *natural* light, clean water, pure air, and open space for relaxation or recreation in contact with nature.[3] The appearance of motorized traffic in the 1920s made the streets as dangerous an environment as the tenement slums in which most of the population lived. Proponents of decent low-income housing, pollution control through zoning, and development of additional parks and green spaces for public recreation could be found in each of these three clusters.

It is true that the widespread criticism of cities neglected very important urban features. The critics disregarded the vitality of working-class culture that developed in the tenement blocks. They paid little attention to the visual stimulation of varied vernacular architecture. They had little to say about cities as economic engines. They overlooked the ways in which urban diversity and conflict could be the source of cultural creativity, even in a mass society. It is also true that very few of the critics, as good intellectuals themselves, were in much of a hurry to flee the city for the unspoiled countryside—except on weekends. Yet there was considerable truth in their charges. Disease and mortality rates far exceeded those in rural areas, and criminality and political radicalism flourished in big cities. In response, middle- and upper-class families were fleeing to the suburbs, especially since they could reach the centers using either new mass transit or their own motor vehicles. In other words, while heritage, preservationist, or Heimatschutz movements waxed eloquent about the historic cities, planners viewed their nostalgia as misplaced sentimentality. The historic but overgrown cities were sick and needed to be healed.

Although some urban centers did sustain damage from artillery and aerial bombing during World War I, in general cities fared better than rural areas. Surprisingly, some Belgian cities suffered greater destruction in World War I than they experienced in World War II. Between 25 and 30 percent of the buildings in Leuven, for example, were destroyed in August 1914. In words that would be echoed by urban planners everywhere twenty-five years later, the architect H. P. Berlage commented in 1919 on a "happy Belgium . . . which is now the first country that will be able to put into practice the clear notion of building a modern town."[4] In fact, Leuven's reconstruction did not so much follow "modern" concepts as "nineteenth-century design typologies," although some modernist plans were put forward.[5]

But probably more important for planners were the demographic catastrophe and economic hardships endured by French, German, and British citizens during the war and the early 1920s. The thought that the deficiencies of big cities would further harm populations already diminished by war made the project of urban reform more urgent. Among reformers the revolutions of 1917–19 generated a euphoric feeling that the time was ripe for new ways of doing things. Together these conditions produced the flowering of the modernist/functionalist models of urban planning.

Consider for a moment the early work of Martin Wagner and Leberecht Migge. Wagner gained fame first as the designer of modern housing projects and then as the chief planner of Berlin in the late 1920s.[6] Already in his dissertation, titled "The Sanitary Greenery of the City" and published in 1915, we can see Wagner's dream of comprehensive planning in the form of "greening" cities. There he dismisses the argument that urban green spaces and wooded parks can act as "green lungs" that will solve the problem of polluted city air. Far too great a forest area would be needed, and medical studies found no statistically significant relationship between the incidence of lung disease and the amount of green space in a specific urban area. City air and air in tenements must be purified by mechanical and chemical means, not by relying on natural green spaces. Open green spaces that are *used* do contribute a great deal to the maintenance of public health because they encourage exercise. Actual use is crucial. If the public can't reach and use urban parks, the mere existence of such spaces is of limited value.

Typical of the aspiration to make urban planning a respected technical profession, Wagner contended that any viable policy on free space must be based on scientific analyses of how these spaces were used and the benefits they provided, such as opportunities for recreation. Wagner sought to quantify the need for green space according to age group, occupation, and location of residence and workplace. He came up with a formula showing that the ideal amount of green space needed for walking and recreation was approximately 19.5 square meters per person. Wagner then conducted a study of forty-six German cities with populations of at least 100,000, which showed that open space in these areas averaged from 2 to 4.3 square meters per person. In England, with figures for twenty-seven cities, the range was 3.6 to 9.1. In cities with more than 500,000 inhabitants, the Germans had only 2.0 square meters, the British 6.1 square meters per person, with corresponding implications for general health. The "solution," Wagner said, "was not in the end a function of the topographical form of the city and its surrounding" but rather of the social, political, economic, and legal conditions of the city.[7] The law needed to be modified to permit land acquisition and maintenance through one-time assessments as well as regular taxation. When granting approval for new developments or improvements of existing properties, it was particularly important that space be set aside for communal use and that this take place without compensation to the property owners. Cities had "a duty to protect the health of the German body public and increase its life force [*Volkskraft*]."[8]

This agenda is quite different from and considerably more ambitious than just building new garden cities or garden suburbs, which at that point was perhaps the main focus of the reform impulse in German planning.[9] The garden-city movement was too limited, too bourgeois, and too modest to accomplish much. Wagner felt that planners had to have the legal tools in order to transform the way *existing* cities were shaped and functioned, with planners working to safeguard the very survival of the people, something made urgently important by Germany's losses in World War I.

An approach related to the issue of urban greenery can be found in the work of the landscape architect Leberecht Migge. Originally known for designing elegant flower gardens for bourgeois clients during the heady days of the 1919 revolution, he wrote under the pseudonym "Spartacus in Green" and, in a series of polemics, promoted another kind of utopian urban vision. Deeply distressed both by the negative impact of "the outdated city of stone" and by the severe economic hardships that resulted from military defeat followed by political turmoil, social dislocation, and rampant inflation, Migge argued that planners must help citizens transform their cities into environmentally and economically sound units.[10]

Like Wagner, Migge stressed the *use* of green spaces, not their romanticized aesthetic character, though in seeing "unused land" as "the land of future youth, of health and happiness," he clearly did evince a romantic streak.[11] He wanted to see cities filled with functioning vegetable gardens, whether truck gardens (*Schrebergärten*), backyard gardens, or actual farms located in green radial spokes or green belts. Such greenery might not eliminate air pollution, but recycling and extensive composting would help to solve urban waste problems, and Migge even produced designs for compost toilets. Intensive private use of communal land, not political revolution, could make individuals and cities self-sufficient and healthy, but cities had to provide the planners with the land they wanted.

Although Migge was awarded a contract in 1922 by the city of Kiel to design a comprehensive plan that would create a complex green belt of forest, agricultural plots, and meadows that would interact with the existing city, the political and economic realities of the 1920s meant that dreams such as Migge's could not be implemented.[12] The same was true for Wagner. He helped to design innovative housing projects that integrated modernist architecture with the environmental concerns of the garden-suburb movement, but by the time he became chief planner of Berlin, the revolutionary energies and sense of opportunity generated by World War I had passed.

The best-known modernist planning model is that most closely associated with Le Corbusier and CIAM. Developed during the 1920s and codified in the 1933 Charter of Athens, the model was supported by architect planners from Britain to Moscow.[13] Although the charter wasn't published until much later, the modernist program was widespread. For example, in 1931 the *Architectural Forum* published an article by Wilhelm Kries, then president of the Association of German Architects (Bund Deutsche Architekten), in which he argues that "the proper way to transform the old dying city districts" was to devote the next thirty years to radical decentralization of population and industry; renovation of run-down areas; and creation of a smoke-free, electrified city with broad streets, superhighways, and open green areas. Indeed, the core ideas of this program were radical not only in their simplicity but also in deriving inspiration from the turmoil that appeared at the end of the Great War. To solve the problem of unhealthy cities—both their historic cores and their sprawling suburbs—one had to embrace the spirit, products (machines), and materials (concrete, steel, and glass) of the modern industrial age. Cities were conceived as having four essential functions: living (housing), working (industry and commerce), recreation (green spaces), and transportation (roads, railways, and airports). Existing cities would be transformed by articulating these functions, which meant clear separation of functions. City centers would be razed and replaced by a thoughtfully designed combination of skyscrapers and open public spaces. Important historic monuments would be preserved, but the "slums" surrounding them would disappear. (Le Corbusier observed at the 1930 CIAM 3 meeting in Brussels that cities of high-rise buildings would be safer from air raids.[14]) Inhabitants of these urban centers could be relocated to residential areas where they would have easy access to green spaces, and factories could be rebuilt downwind in order to minimize exposure to various forms of pollution. Industries and housing would be located so as to minimize commuting. Whether building garden suburbs of low-rise houses or tall, slablike apartment blocks, the built environment would reflect the natural landscape, topography, and climate. It was this reflection that would make each city unique, even though the fundamental principles were universal. The modernist model paid relatively little attention to the cultural, aesthetic, or political functions of historic city centers—the things that in fact had also contributed to the unique character of each city. CIAM members busily prepared plans to renovate cities like Berlin and London as if they could sweep the existing cityscape clean and start over with blank slates—projects that seemed quite utopian.

Another kind of utopian planning in the 1930s was that of German architects working under the auspices of the National Socialist government to create monumental buildings and avenues in some eighty cities.[15] Albert Speer's plans for Berlin are but the most famous. Although the formulaic, monumental neoclassicism of the designs had little in common with the tower slabs of CIAM, the German architects, too, intended to modernize their cities by, for example, providing broad streets not just for political rallies but also for motor traffic. Begun in 1937, these planning projects continued into the war years, though actual implementation was suspended because war projects enjoyed higher priority. Moreover, the cost of implementation would have been so enormous that only a complete military victory would have produced the required resources.

World War II, which from the very beginning brought vast destruction to cities, changed the situation for urban planners in some ways but not in others. Since World War I, planners had been aware of the horrible potential of aerial bombardments, which dramatically transformed the urban environment: high explosives reduced brick-and-mortar buildings to rubble, while fire turned wooden structures into ashes. Cityscapes looked more and more like surreal natural landscapes—wild moonscapes littered with ruins that would have appealed to the imagination of nineteenth-century romantics and indeed held a fascination for everyone who saw them, including embattled and impoverished residents and the Allied soldiers who would later occupy them. As World War II dragged on, surviving trees on boulevards and in parks were chopped down for fuel, further denuding the urban landscape. While most of the destruction was above ground, water and sewer lines were also damaged by bombs. Rivers and canals were blocked with barges, and fuel spilled into the waters.

One significant reaction of planners was in the area of civil defense. Densely built-up cities were clearly susceptible to spreading fires, and narrow, clogged streets made it difficult not only for firefighters to respond but also for residents to be evacuated. An obvious, though partial, solution was to cut firebreaks in the form of broad avenues and to relocate the inhabitants of heavily populated neighborhoods to safer suburbs or the countryside. Planners would not have considered this approach incompatible with prewar ideas about modernizing cities. Early in the war and in those countries under attack, there was of course no opportunity to execute such massive projects. Elsewhere there were important economic, political, and psychological obstacles. Compensation would have to be made to property owners whose buildings were demolished in order to create firebreaks. The government would have to prepare citizens for the likelihood of aerial bombing and, in so doing, implicitly acknowledge the possibility of ultimate defeat.

The Japanese, both remembering the 1923 Tokyo earthquake and aware of the devastation of European cities by Allied bombing (which did not begin in Japan until quite late in the war), passed a law in 1940 that authorized municipal governments to create firebreaks. In several cities, including Nagoya and Tokyo, firebreaks were constructed as a preventive measure, with more than

610,000 houses being demolished in the process. Drastic as this action was, it did prevent some neighborhoods from being destroyed in the fires started by U.S. bombing. In the postwar period these firebreaks became important transportation routes.[16]

Instead of building firebreaks, the Germans concentrated on forced evacuations to the countryside and the construction of subterranean and aboveground bunkers. The latter made a dramatic impact on the cityscape, as one can see in the photos published by Jörg Friedrich in his recent book *Brandstätten*.[17] Sometimes erected in parks, sometimes in densely built-up areas, these bunkers did not always provide the promised protection, but they remained standing when buildings around them were reduced to rubble. Indeed, they proved costly and sometimes very difficult to remove after the war and remained visible if grim landmarks for years to come.

There was, however, at least one interesting exercise in urban planning that focused primarily on air raids. In 1943, after Cologne, Hamburg, and other cities had been badly damaged, Gerhard Graubner wrote an essay titled "The Idea of War as the Basis of City Form and City Planning." Graubner argued that in rebuilding the bombed cities, planners must make the protection of inhabitants and preservation of the city's key infrastructure their top priorities. This would also be an opportunity to reconnect or reintegrate the city into the natural landscape. Currently massive aboveground bunkers were being built in the middle of cities, but these were not successfully harmonized with existing styles and structures, even if some had been planned with the intention of masking them with some sort of external cladding and eventually converting them into garages, warehouses, or whatever.[18]

According to Graubner, the policy of mass evacuation was misguided. The city ought to be a place where rural populations could take refuge rather than one from which urban populations fled. Nor was there any need to relocate industries to the countryside in order to ensure access to resources such as coal and iron ore, since rail lines damaged in air raids could be repaired relatively easily, as past experience had shown. Instead, he proposed rebuilding cities in terrains that offered natural protection from air attacks, and constructing all main transportation routes, both roads and railways, underground, reserving "above ground streets for local traffic, bicycles, and pedestrians."[19] Utility lines would also run underground. Tower bunkers in the city center would provide shelter in emergencies and be linked to underground transit. Surrounded by green spaces, these towers could include housing units, although most residents would live in satellite communities of one- and two-story buildings separated from one another and from the city center by public green spaces, including truck gardens and recreational areas. Ideally the main administrative, commercial, and cultural buildings would be in the city center with housing to the west and south and industry located in the north and east.

A perfect example of a city where this approach might work was Stuttgart, which was situated in a kind of three-sided bowl. Hills were already there for tunnels to serve transit and industry. Flak batteries could be mounted on the hilltops without disturbing the core city. In the case of Hannover, where industry was currently located to the city's west and southwest, this scheme would mean relocating housing to the west of the industrial area, where the Gehrdenerberg and Benterberg would provide appropriate air-raid protection. Connections between the residential, industrial, and commercial areas would require a new, bomb-proof underground train station and rail lines. Broad green areas would help to reduce the density of the historic city, which had already lost its historic character in the bombing. Graubner's is a remarkable vision—the city as permanent bomb shelter!

Note that this combination of modernist design with principles of civil defense had nothing to do with the neoclassical monumentalism typical of German planning right before the war. This was also true of three other kinds of German wartime planning. The first dealt with planning new towns, the second with planning for areas conquered and occupied by German troops, and the third with planning postwar reconstruction of cities destroyed in the conflict.

While some planners continued to devote attention to the projects for monumental avenues and buildings for cities within prewar Germany, actual construction was not possible during the war. Hence they also worked on models for new towns. In 1940, Konstanty Gutschow, Wilhelm Wortmann, and Hans Bernhard Reichow worked on articulating two key concepts, the *Ortsgruppe als Siedlungszelle* and *Stadtlandschaft*.[20] The former was a concept of relatively small residential settlements, or cells, of five to eight thousand inhabitants. They would be surrounded by green areas, maximizing light and fresh air but also reducing possible danger from air raids. The cells would be clustered around central community buildings and, taken together, would form a basic political unit (*Ortsgruppe*) in the new Reich. *Stadtlandschaft* meant embedding new cities or extensions of existing cities into the landscape, whereby cityscape and natural landscape would become one. Here the modernist grids were rejected in favor of organic forms that followed rivers and the local topography.

Within the sheltered confines of the German Academy for City, Reich, and Regional Planning (this organization was also involved in planning for the conquered lands in the East), Johannes Göderitz, Roland Rainer, and Hubert Hoffmann elaborated a model for both new cities and new suburban settlements that was very close to the modernist/functionalist model promoted by CIAM. This was not entirely a surprise, since Hoffmann in particular had been close to that movement, and he would be the first to publish a German translation of the Charter of Athens after the war. The work of Göderitz and his colleagues remained a planning exercise and wasn't published until 1957, but it was already widely circulated and thus well known before the war ended.[21] It was enormously influential in guiding the thinking of planners during postwar reconstruction.

Town plans for Alsace and western Poland, both annexed to Germany, and what was called the "East" (former Polish and Soviet territory), derived from an odd combination of sources. The war made possible a blend of environmental concerns, epidemiology, racism, eugenics, settlement planning, and outright programs of genocide. Here the Germans were talking not so much about the reconstruction of war-damaged cities as about building a network of small new towns that were to be the opposite of unhealthy industrial cities. They drew ideas from Theodor Fritsch's 1896 *Die Stadt der Zukunft* (*Gartenstadt*) [*The City of the Future: Garden City*] and Gottfried Feder's *Die neue Stadt* [*The New City*], published in 1939. A notorious anti-Semite, Fritsch had advocated redistributing the population of big cities into new garden cities, where communal ownership of the land and long-term leases would protect the Aryan Germans from the Jews who controlled both land and money in the cities. These garden cities would produce a healthy German race.[22] Gottfried Feder, one of the original members of the National Socialist German Workers' Party (NSDAP), enjoyed brief prominence between 1933 and 1939 as a professor of planning at the Technical University of Berlin. A life-long anti-Semite, in his 1939 book he advocated the creation of new, twenty-thousand-inhabitant towns that were in some respects modeled on the garden-city ideas of both Fritsch and Ebenezer Howard, founder of the British garden-city movement.[23] Although the morphology of Feder's model cities resembled medieval towns more than the modernist's functional city, they were nonetheless to be sited so as to benefit from the natural advantages of the locale. Once the Germans had acquired their new eastern territories, there was something of a struggle between Walter Darré, Reichsminister of Food and Agriculture, and Heinrich Himmler, head of the SS, over the character of future settlements. Darré wanted purely agrarian settlements; Himmler wanted to use the creation of new towns as part of an aggressive demographic and political program to Germanize and transform Eastern Europe, by which a million Germans or people of German descent would oversee an enslaved population of non-Germans. Himmler won out. In late January 1942 (only weeks after the Germans failed in their attempt to capture Moscow), Himmler issued a planning ordinance titled Guidelines for the Planning and Design of Cities in the Annexed German Territories in the East. It called for an "organic" network of cities with fifteen to twenty thousand inhabitants. Designed to reflect "the spirit and needs of the present time," these towns would harmonize "with the natural forms of the landscape." Residents would live in "cells" of about five thousand, which would also constitute the basic political unit. There would be plenty of green areas; industry would be located so as to minimize environmental harm, and heavy traffic would be routed around the towns rather than through them, thus giving pedestrians and bicycles priority in the city centers.[24] Konrad Meyer, the head of the SS planning department, incorporated these ideas in May 1942 into what became known as the Generalplan Ost. It called for building a matrix of towns, more or less like those proposed by Feder. The matrix would be based on the "central place theory" developed by Walter Christaller and the regional planning concepts developed within various institutions responsible for regional planning, including the Reich Office for Regional Planning, created in 1935, and the German Academy for City, Reich, and Regional Planning. From the planners' perspective, in other words, the war presented an unprecedented opportunity to design a vast territory to their (and Himmler's) liking.[25] These ideas, developed during the war, helped to form the basis for regional planning undertaken after 1945.

The sense that war meant great opportunities for planners did not change at all when allied air raids began to destroy German cities. Recall Berlage's remark in 1919 that the destruction of Leuven and the opportunity to build a modern city should make Belgians "happy." No sooner had bombs fallen on Lübeck, Hamburg, and other cities than local planners and those involved in the monumental redesign projects began to draft plans for rebuilding. In December 1943 this activity was organized by the task force for reconstruction planning (known as the Arbeitsstab Wiederaufbauplanung bombenzerstörter Städte) within Albert Speer's armaments ministry. Here, too, the emphasis was on the functional aspects of modern cities, not on architectural form.[26]

This task force was coordinated by Speer's office manager, Rudolf Wolters, himself a planner, and Konstanty Gutschow, the architect chosen for the redesign of Hamburg. The group held a series of meetings at Wriezen near Berlin between August and November 1944, where it discussed a wide variety of subjects.[27] These topics included, among others, the creation of standards for measuring war damage; issues, methods, guiding principles, and standards to use in reconstruction planning; and the expected economic burdens of rebuilding. Unconcerned with aesthetic matters, these architect/planners were at this point interested neither in the loss of historic architecture nor in the monumental building programs laid out for most large cities before the war. Instead their focus was on practical matters, such as housing, green spaces, transportation networks, and the like. Rebuilding was seen as an opportunity to make improvements in the urban fabric. Most of those who took part in these discussions—Wolters, Gutschow, Ernst Neufert, Hans Stephan, and Friedrich Tamms—went on to play important roles as planners or architects after the war.

The protocols of the Wriezen meetings reveal a number of ways in which the war and its impact influenced thinking about rebuilding. For example, Gutschow, in a general building plan for Hamburg, writes that one must not be utopian in rebuilding Hamburg, even if that means accepting still-standing areas in need of renewal that had been breeding grounds for communism. The rebuilt cities ought not to consist "of amorphous masses of dwellings" but rather be built around "individual cells of lively settlements."[28] Reinhold Niemeyer declares that rebuilding should be embedded in *Raumordnung*, or regional planning, that considers: (a) the natural topography and climate; (b) features created by humans, including transportation, trade and industry, agriculture; and (c) race, settlements, housing, and other aspects of state political goals. Above all, however, planners must consider topographical features, including river valleys, mountains and hills, soil, and climate. "Fundamentally the natural economy (*Haushalt der Natur*) in a region must be kept in balance as much as possible."[29] Water runoff, flows of treated domestic sewage, and flood-prevention measures must be carefully negotiated and brought into balance, with special attention given to supplies of water for drinking but also for hydroelectric energy production.

According to Willi Schelkes, reconstruction should be based on the concepts of work, living (housing), and recreation. In terms that could have been used by Martin Wagner during World War I, Schelkes asserted that space for recreation was more important than just free or open space or town forests because recreation improved public health. City-building over the preceding century had eliminated most recreational free spaces, and reconstruction was an opportunity to reverse this process. Schelkes saw recreational spaces as so essential to public health that he considered it "better to build and maintain green areas than to build hospitals."[30] Furthermore, broad, tree-lined streets and large squares—especially if laid out as spokes from the center—could act as effective firebreaks in the event of future air raids.[31]

Preparing cities for future wars was often on the minds of the planners during their discussions in Wriezen. Hans Stephan and Fritz Leonhardt spoke of the need to use steel and concrete wherever possible, including in housing construction, because those materials better resisted fire damage.[32] The use of modern materials, plus heavier insulation, would also improve heating efficiency. (After the war the Americans urged the Germans to build wood-frame houses and did not understand why they preferred not to use highly flammable materials, even if doing so would lower the cost of construction.) Herbert Rimpl, who before the war had designed the new industrial city for the Hermann Göring steel works (now Salzgitter) along functionalist lines, stressed that when rebuilding industrial cities, the location of factories should be determined not only according to wind direction but also by considering which sites offered the best protection from aerial attack.[33]

The planners in this task force for reconstruction planning prepared detailed plans for a number of cities and distributed memoranda containing recommended standards and methods to planning offices throughout Germany. As I have argued elsewhere, few of these men were able to implement their plans directly after the war, but the models, standards, and procedures, combined with the models developed by Göderitz and his colleagues, formed the basic framework within which much postwar West German reconstruction took place. When the war ended, there was a cohort of well-prepared planners on hand, eager to apply their ideas and experiences. Viewed as technocrats, very few of these men saw their careers hurt by denazification. Only the Nazi-era vocabulary and insistence on the political character of settlement cells had to be dropped. Monumentalism in architecture also disappeared with the Third Reich. Rebuilding was seen as a great opportunity to reshape

Germany's cities for the better, and the predominant model was a kind of modernist functionalism.

It is important to realize that this model was also to be found in France, Britain, and even the United States during the war. In these countries, too, the circumstances of the war appeared to planners as a unique opportunity to transform the urban landscape. The fact that no American, few French, and only some British cities were actually damaged in air raids suggests that "circumstances" included not just destruction but the sense of emergency, the widespread emphasis on the importance of planning of all kinds.

It is possible to give here only a few indications of the scope of wartime urban planning in these countries, so a few examples must suffice. In France, Le Corbusier first hoped that he could realize his prewar plans to transform the city of Algiers into a modern city of the future. When that became impossible, he sought to work for the Vichy regime, and although he fell out of favor, he attempted to realize versions of his utopian "radiant city" in Saint-Dié and La Rochelle–La Pallice. While none of his plans were executed until after the war, he did publish the Charter of Athens in 1943. There and in other publications he continued to insist on the absolute centrality to modern urbanism of air, sunlight, greenery, and open space.[34] While Le Corbusier himself played no significant role in actual postwar rebuilding, his ideas influenced the work of others, including August Perret, the architect responsible for rebuilding the port of Le Havre. At the same time, reconstruction planning in the bombed cities of the Loire Valley was based on an "aesthetic compromise" between a traditional reverence for local architectural traditions and functionalist ideas about light, space, and broad streets for traffic.[35] Again, many planners saw the devastation wrought by aerial bombing as an opportunity to modernize the urban landscape.

The war years also saw an extraordinary outpouring of town plans in Britain. Peter Larkham and Keith Lilley have identified 195 reconstruction plans for 131 places that were prepared between 1939 and 1952. Many of these were for cities that had suffered bomb damage, but plans were also prepared for several undamaged towns where administrators believed both that the time for comprehensive planning had arrived and that resources would be forthcoming.[36] Some plans were modest and conservative, but many others were radical, "wiping the slate clean even if bomb damage had not provided the opportunity to do so," to quote Larkham's and Lilley's assessment.[37] Donald Gibson's modernist design for reconstructing the bombed center of Coventry, for example, was inspired by the Modern Architecture Research group (MARS), an organization of architects affiliated with CIAM that had been working on radical plans for London throughout the 1930s. These drawings show new, broader streets and squares, with buildings either clearly modern in form or perhaps modern in structure even though constructed of traditional brick.[38]

The desire to seize on the war as a chance to reshape cities also spread to the United States. While it is important to realize that architecture and town planning were international professions, with participants reading about one another's work, meeting at conferences, corresponding, and traveling to see things firsthand, it is also true that the functionalist/modernist model was carried to the United States by European émigrés. Bauhaus founder Walter Gropius and Martin Wagner, both of whom had participated in the CIAM Brussels meeting of 1930, accepted posts at Harvard after leaving Germany. Between 1941 and 1943, Gropius and Wagner coauthored a half-dozen essays and reports on city planning and housing in wartime America. Wagner considered the condition of blighted U.S. cities, filled with overcrowded slums, comparable to that of Europe's bombed cities. The wave of war-related planning created enthusiasm in America for demolishing older urban areas and rebuilding them according to the new the functionalist model.

One piece by Wagner and Gropius, titled "Cities' Renaissance," begins with the exclamation "Our cities are sick, deathly sick, machine sick!"[39] American cities built by the railroad age were decaying. Now was the time to build "self-contained town-cells or townships," where living would be on a human scale.[40] Skyscrapers in cities like New York, they said, "have become symbols of light robbers, traffic compressors, and space squeezers; people begin to hate being poured daily into stone masses and crematoriums of real life and happiness."[41] What was needed instead was large-scale comprehensive planning, national building laws, and "compulsory amortization and depreciation of all building structures" so that planning authorities could demolish and renew.[42]

Postulating that people flee the cities for the suburbs in order to live closer to nature, Wagner and Gropius argued that planners should devise a transportation system to ease traffic flow between the old cities (which should be rejuvenated) and the new neighborhoods on their peripheries. Hence "the goal of the modern town-planner is to bring town and country into a closer and closer relationship."[43] The excess population of the old cities should be siphoned off to populate the new small towns, freeing up space in the cities for renovation. Tracts of city land should be acquired by the public and consolidated to facilitate new planning and development. In this way, "All those employed in the central areas will live in dwelling quarters which, more widely spaced and surrounded by parks, will fit their inhabitants for the role of building that constructive community interest and neighborhood spirit long lost in the old cities."[44]

Other European architect/planners now in America also contributed to the spread of the modernist planning model. José Luis Sert, a Spanish architect now living in New York who had been an active CIAM participant, published a book in 1942 that served as an American version of the CIAM manifesto. Titled *Can Our Cities Survive? An ABC of Urban Problems, Their Analysis, Their Solutions: Based on the Proposals Formulated by the C.I.A.M*, this work outlined not only the need to renovate cities but also the ways in which plans on the CIAM model would help to protect American cities from air raids. Thus earlier efforts at urban renewal replaced slums with other high-density structures when they should have left open spaces and created air-raid shelters. Congested cities "offer excellent targets from above." The most effective and economical approach is housing people in "high buildings separated by large spaces" in which bomb shelters could be built. Targets would thus be more difficult to hit, and people could be more quickly moved to underground shelters. Moreover, the possibility of air raids encouraged the redistribution of industry and its separation from residential areas.[45]

To sum up, then, planners from all over Europe and America—on both sides of the conflict—saw World War II as presenting a unique opportunity to rebuild, renovate, redesign, and rejuvenate older cities blighted either by damage from aerial bombing and field artillery or by unplanned development and other urban plagues. As cities were rebuilt, new towns or settlements would be created according to modern ideas of the functional city. The built environment, the cityscapes, would be embedded in the natural landscape or, at the least, brought into contact with natural light, air, and greenery. This task was urgent if cities, their peoples, and, by extension, their nations were to survive and flourish.

Is this what happened when the war ended? Recent works by the historian Jörg Friedrich and the late novelist W. G. Sebald dealing with the bombing of Germany have unleashed a flood of attention to how Germans coped with the ruin of their cities. Sebald said that "the destruction, on a scale without historical precedent, . . . has been largely obliterated from the retrospective understanding of those affected." Sebald faulted the Germans for ignoring the destruction and instead undertaking to build "a brave new world."[46] Friedrich described the air war as "an inexplicable orgy of destruction [*Vernichtungstrunkenheit*]," noting that it had "obliterated the image that the cityscape [*Stadtlandschaft*] had produced from the historic landscape"; and he lamented the fact that his countrymen had made no attempt to "restore the unity of space and history" but instead had turned their backs on the past.[47] For Friedrich, Germany's cities were old and beautiful and now, sadly, gone forever.

This view of German cities—and many non-German cities—was clearly *not* shared by town planners. Not everything had been beautiful and valuable in the crowded, unhealthy centers of big cities. Nor did planners ignore the destruction; rather, they catalogued it in great detail. They wanted to build a world of modern cities based not on amnesia about the destruction but on what these planners considered progressive, scientific knowledge about the structural and functional needs of cities. The prewar and wartime dreams of transforming cities persisted in the postwar era.[48]

The damage was of course enormous, especially in Japan and Germany, but also in Poland and parts of the USSR. Depending on whether a city had been built of wood or stone, what kinds of weaponry had been used, the frequency and number of raids or duration of artillery sieges, damage to the built environment ranged from 3 percent in British towns to 95 percent in towns on the continent. Nearly two million tons of bombs had been dropped on Axis

Europe. Of this, 31 percent had fallen in so-called area raids on cities. Of German cities with more than 100,000 inhabitants in 1939, on average about 50 percent of their built-up areas were more or less destroyed. In Würzburg the figure was 89 percent, in Remscheid and Bochum 83 percent, in Hamburg and Wuppertal 75 percent. Some 45 percent of the residential housing in large cities was destroyed. Comparable numbers of schools, churches, town halls, banks, factories, and shops were also in ruins.

Obviously, buildings and utilities were not the only casualties. Human casualties were staggering. Perhaps 35,000 died in Hamburg during a single night raid in mid-1943, and about the same number died in Dresden in 1945. Somewhere between 400,000 and 600,000 civilians died in the air raids; between 650,000 and 850,000 were injured; and millions were made homeless or were evacuated to the countryside in anticipation of bombing. But the main impression in 1945 was of ruins that remained, not of casualties that had long been buried.

It was estimated that at war's end Berlin contained 55 million cubic meters of rubble, Hamburg 36 million, Cologne and Dresden around 24 to 25 million, Dortmund 16.8 million, Nuremberg 10.7 million, Bremen 8 million cubic meters, and so on. Officials in Munich noted that their 5 million cubic meters of debris equaled double the mass of Egypt's Great Pyramid.[49] In Hamburg it was observed that if that city's rubble were to be loaded into normal railroad freight cars, the train would be long enough to circle the Earth.

The chief planner of Lübeck expected rebuilding to take from sixty to eighty years. The destruction of a few German cities was so great that authorities proposed abandoning them entirely and building new cities, perhaps underground, to ensure protection from air raids. Other proposals advocated bulldozing the rubble, planting grass and shrubbery, and allowing a natural landscape to emerge. A new city could then be built that would harmonize with the landscape. A famous proposal for rebuilding Berlin advocated such an approach: Berlin would have emerged as a kind of ribbon city following the course of the Spree. Had plans like this been implemented in Germany's bombed cities, the first step would have been to relocate the remaining population to the countryside. If executed in every bombed city, this program would have returned Germany to a pastoral state. There were similar proposals for rebuilding Hiroshima on a new site, thus abandoning the area hit by the atomic bomb.[50]

In fact, as we know, no bombed cities were abandoned. All were rebuilt, and remarkably quickly. To what extent did the planners succeed in creating a new cityscape or urban landscape, especially one along the lines called for by the modernists? Were the rebuilt cities or the unbombed but blighted cities filled with light, air, and greenery? Were they embedded in the natural landscape? In short, were planners able to seize this opportunity to build new cities or modernize existing cities?

The answer, alas, is a frustrating yes and no. The results of postwar planning in the vast area from Japan to America to Europe to the Soviet heartland are tremendously varied and defy easy generalization. For that reason it has been a more manageable task for scholars to examine plans and planners rather than the messy cities that were built.

The rubble was of course cleared away, and its removal often changed the topography: the postwar moonscapes disappeared. Some rubble was salvaged or reprocessed into new building material; most went into fill. Cellars everywhere were filled, often by property owners anxious to get started with rebuilding. Hamburg used rubble to fill in canals. Munich, Cologne, and Berlin used it to create new parklands, all of which became important recreational areas for future residents. Rubble in Rotterdam was used to fill up a river. In central Le Havre it was used to raise the ground level by almost a meter, which "freed the planners from the ancient layout" and made it possible to place all utilities underground.[51] The rubble of the Warsaw ghetto was used to raise the level of the blocks in that area so that rebuilt structures could stand on terraced land. In Japan, with its mostly wooden buildings, there was not the volume of rubble as in Europe; but in Osaka, planners used the opportunity created by reconstruction to raise the level of the harbor area.[52]

Before the war, planners had fantasized about having urban blank slates on which to plan and build. The vast war damage appeared to have actually created those blank slates. Everywhere, however, modernist-leaning planners found the path to fulfilling their dreams blocked by formidable obstacles. The very practical things that they had ignored in their earlier manifestos—how to acquire and consolidate large parcels of land for redevelopment, how to resettle large numbers of people, how to incorporate important pieces of historic architecture—hindered radical reconstruction.

In Western Europe, Japan, and the United States, property owners fought tenaciously against the efforts of public authorities to determine whether and how they might rebuild, and planning offices seldom had the legal authority to impose major reconstruction plans on the citizenry. Illegal or "wild" rebuilding went on everywhere, despite the efforts of building inspectors, and many "temporary" structures became permanent. Having endured years in which worldwide economic depression and two world wars gave governments extraordinary authority, peacetime and the return (or turn) to democratic government made it hard for town planners to run roughshod over private citizens and private businesses.

Frequently there were conflicts among local, regional, and central authorities over control of the rebuilding process, and everywhere plans were undercut by the lack of financial resources. This undermined reconstruction planning in Britain and Japan and also in Eastern Europe. The central ministries in Britain and Japan either blocked or overrode radical local initiatives and declined to provide funding. Thus Gibson's plans for Coventry, which was to be the test case for a modern, rebuilt Britain, were only partly realized. Charles Holden and William Holford were unable to transform central London. The new laws governing plot ratios to increase access to natural light that they helped to introduce did not significantly reduce densities or resolve traffic congestion. Ishikawa Hideaki's radical plans for decentralizing and greening Tokyo failed to get the support of the national government and were shelved.[53] In Eastern Europe, central authorities blocked local initiatives, but in some cases town citizens managed to obstruct plans of the central government. In Dresden, local planners, political figures, and citizens resisted the attempts of Berlin to impose Stalinist models, though they were unable to prevent the wholesale clearing of most of the central city. Elsewhere the central authorities prevailed, as in East Berlin, Königsberg (Kaliningrad), Sevastopol, Leningrad, and Stalingrad, often with dreary results. Soviet-bloc models of representative "socialist" cities turned out to be awkward combinations of monumentalism and modernist functionalism.

While planners in the West were focused on modernizing cities, they usually had to accede to the wishes of preservationists, church authorities, and civic leaders that important historic monuments from earlier centuries be rebuilt.[54] Historic buildings, restored at least in their external form, again helped to define cityscapes, although very few cities were rebuilt as replicas of what had stood before the war. In West Germany, Rothenburg ob der Tauber was rebuilt in its prewar form, as, for the most part, was Freudenstadt, but both were quite small. In rebuilding Warsaw and Gdansk, Communist authorities confiscated private property, redrew building-plot lines, and modernized the interiors of both blocks and individual buildings; but an effort was made in the city centers to replicate the prewar exteriors of many buildings.[55] Outside the "historic" centers of Warsaw and Gdansk, however, the layouts of the rebuilt cities follow modern, functionalist lines.

But faithful replication of the past was as much an exception as realization of the radical dreams of the modernist planners was. Much more common were compromises between what most planners dreamed of, what property owners and heritage-minded citizens wanted, and what was possible, given the scarcity of resources. In West Germany, the Netherlands, France, and Britain postwar planning models were predominantly modernist/functionalist, however little or much of them was actually realized. The urban landscapes of the rebuilt cities mix together the planned and unplanned. Restored historic buildings stand alongside starkly modern structures that are functional modern but seek to adapt to local styles by using traditional external materials or approximating traditional roof lines and building proportions. In the postwar era, planners built broad new boulevards and widened other city streets, such as the hundred-meter-wide arteries in Hiroshima and Nagoya, the Thälmannstraße and Prager Straße in Dresden, the Stalinallee in East Berlin, and the North-South Street in Cologne. At the same time, major parts of the prewar street patterns persisted everywhere, often because of the need to reuse underground utilities. Population densities in the city centers were reduced, partly through the efforts of planners but also because of continued flight to planned garden suburbs or new towns alongside uncontrolled sprawl. Japanese cities were rebuilt in a helter-skelter fashion that deeply disappointed Japan's planners.

Early in this chapter I pointed to long-term continuities in town-planning models: the garden city, *Stadtlandschaft,* the modernist/functionalist model. If these models persisted over time, we must ask how much it mattered to the urban landscape that cities were bombed in wartime. To what extent was the urban environment (the built environment in its natural and topographical setting) different as a result of war?

When we look at American cities, or even at undamaged European cities in the postwar period, we see that most of them were also modernized. Huge changes were introduced not through rebuilding war-damaged areas but through urban renewal, the construction of urban highways and other transportation arteries, the building of apartment tower blocks, and the relocation or closing of antiquated smokestack industries. Boston, Basel, Kyoto, and Brussels—to mention but four unbombed cities—do not look today as they looked in 1939, any more than do Berlin, Rotterdam, or Le Havre. But where bombs fell, there was more and more rapid change. For better or worse, bombed cities became the laboratories for postwar urban design. It was not until the late 1960s that planners turned away from the models that had so long prevailed in order to embrace new models based on population density rather than dispersal, multiple rather than solitary functions, pedestrian rather than motorized traffic, and urbanity rather than a general critique of the city.

Notes

1. Martin V. Melosi, "The Place of the City in Environmental History," *Environmental History Review* 14, no. 1 (1993): 1–23; Christine M. Rosen and Joel A. Tarr, "The Importance of an Urban Perspective in Environmental History," *Journal of Urban History* 20 (1994): 299–310; and the essays in Christoph Bernhardt, ed., *Environmental Problems in European Cities in the 19th and 20th Century/Umweltprobleme in europäischen Städten des 19. Und 20. Jahrhunderts,* Cottbuser Studien zur Geschichte von Technik, Arbeit, und Umwelt 14 (Münster and New York: Waxmann, 2001). These essays were originally presented at the Fourth International Conference on Urban History (Venice, 2000). Joel Tarr's essay therein,"Urban History and Environmental History in the United States: Complementary and Overlapping Fields," reinforces the argument made in his 1994 essay.

2. See the essays in Rainer Hudemann and François Walter, eds., *Villes et Guerres mondiales en Europe au XXe siècle* (Paris: Harmattan, 1997); and the January 2003 issue of *Informationen zur modernen Stadtgeschichte,* ed. Dieter Schott, which has as its theme "Stadt und Katastrophe" [City and Catastrophe]. In August 2004 the Centre for Metropolitan History at the Institute of Historical Research in London sponsored a major conference on "Metropolitan Catastrophes: Scenarios, Experiences, and Commemorations in the Era of Total War," but according to Stefan Goebel, the organizer, none of the papers dealt with environmental issues.

3. See, for example, Reinhard Spree, *Soziale Ungleichheit vor Krankheit und Tod. Zur Soz-*

ialgeschichte des Gesundheitsbereichs im Deutschen Kaiserreich (Göttingen: Vandenhoeck & Ruprecht, 1981); and Klaus Bergmann, *Agrarromantik und Großstadtfeindschaft*, Marburger Abhandlungen zur Politischen Wissenschaft 20 (Meisenheim am Glan: A. Hain, 1970).

4. Quoted in Marcel Smets, "The Reconstruction of Leuven after the Events of 1914," in *Villes en mutation XIXe–XXe siècles*, Collection Communal de Belgique, no. 64 (Brussels: Crédit Communal del Belgique, 1982), 501. My translation of this sentence differs slightly from Smets's.

5. Ibid., 504–5. See also the essays in *Resurgam: La reconstruction en Belgique après 1914*, ed. Marcel Smets (Brussels: Crédit communal, 1985); and Pieter Uyttenhove, "Continuities in Belgian Wartime Reconstruction Planning," in *Rebuilding Europe's Bombed Cities*, ed. Jeffry M. Diefendorf (London: Macmillan, 1990).

6. For a biographical sketch of Wagner, see Klaus Homann, "Biographie, Werkverzeichnis, Bibliographie," in *Martin Wagner 1885–1957. Wohnungsbau und Weltstadtplanung. Die Rationalisierung des Glücks* (Berlin: Akademie der Künste, 1985), 157–87.

7. Martin Wagner, *Das sanitäre Grün der Städte. Ein Beitrag zur Freiflächentheorie.* (PhD diss., Technische Universität Berlin, 1915), 34–35.

8. Ibid., 92.

9. The most famous German garden city was Hellerau near Dresden. The work of Fritz Schumacher, who created green belts in Cologne with the support of then-mayor Konrad Adenauer, has much in common with Wagner's vision as well as that of the garden-city movement. Schumacher also planned new roads to cut through the inner city and new fringe settlements. See Diefendorf, "Städtebauliche Traditionen und der Wiederaufbau von Köln vornehmlich nach 1945," *Rheinische Vierteljahrsblätter* 55 (1991): 252–73; and Schumacher, *Köln. Entwicklungsfragen einer Großstadt* (Cologne: Saaleck Verlag, 1923).

10. Fachbereich Stadt- und Landschaftsplanung der Gesamthochschule Kassel, *Leberecht Migge, 1881–1935. Gartenkultur des 20. Jahrhunderts* (Bremen: Worpsweder Verlag, 1981). Quote from Migge's "Das Grüne Manifest" of 1919, here 13. See also the essay by Jürgen H. von Reuß, "Gartenstadt als Stadt der 'Gärten': Leberecht Migges Beitrag zur Gartenstadt," in Franziska Bollerey, Gerhard Fehl, and Kristiana Hartmann, eds., *Im Grünen wohnen — im Blauen planen. Ein Lesebuch zur Gartenstadt mit Beiträgen und Zeitdokumenten* (Hamburg: Christians, 1990), 247ff.

11. Reuß, "Gartenstadt als Stadt der 'Gärten.'"

12. See the drawings in Fachbereich Stadt- und Landschaftsplanung, 105. Jürgen von Reuß, in "Leberecht Migge — Spartakus in Grüne. Das konsequente Experiment des Sonnenhofs in Worpswede," notes that the police saw Migge as a Communist, while the Left viewed him as an odd kind of reactionary. He had an abrasive personality that irritated colleagues, many of whom welcomed his decision to resign from the Bund Deutscher Gartenarchitekten in 1929. Ibid., 11–12.

13. A good recent summary can be found in Eric Mumford, *The CIAM Discourse on Urbanism, 1928–1960* (Cambridge, MA: MIT Press, 2000). See also Kenneth Frampton, *Le Corbusier* (London: Thames & Hudson, 2001), 46–54.

14. Mumford, *CIAM Discourse*, 49.

15. Some of these plans are discussed in Jeffry M. Diefendorf, *In the Wake of War: The Reconstruction of German Cities after World War II* (New York: Oxford University Press, 1993).

16. See Diefendorf, "War and Reconstruction in Germany and Japan," 213; and essays

on various Japanese cities in Carola Hein, Jeffry M. Diefendorf, and Ishida Yorifusa, eds., *Rebuilding Urban Japan after 1945*, (London: Palgrave Macmillan, 2003).

17. Jörg Friedrich, *Brandstätten. Der Anblick des Bombenkriegs* (Berlin: Propyläen, 2003).

18. Gerhad Graubner, "Der Wehrgedanke als Grundlage der Stadtgestaltung und Stadtplanung," reproduced in Werner Durth and Niels Gutschow, *Träume in Trümmern. Planungen zum Wiederaufbau zerstörter Städte im Westen Deutschlands 1940–1950*, 2 vols. (Braunschweig and Wiesbaden: Friedrick Vieweg & Sohn, 1988), 2:771–76.

19. Ibid., 773.

20. For these concepts, see Diefendorf, *In the Wake of War*, 165–66. See also Werner Durth, "Die Stadtlandschaft. Zum Leitbild der gegliederten und aufgelockerten Stadt," in *Architektur und Städtebau der fünfziger Jahre*, ed. Werner Durth and Niels Gutschow Schriftenreihe des Deutschen Nationalkomitees für Denkmalschutz, vol. 41 (Bonn: Das Nationalkomitee, 1990). Two recent treatments of the concept of *Stadtlandschaft* are Elke Sohn, "Hans Bernhard Reichow and the Concept of Stadtlandschaft in German Planning," and Panos Mantziaras, "Rudolf Schwarz and the Concept of Stadtlandschaft," both in *Planning Perspectives* 18, no. 2 (2003): 119–46 and 147–76. A rather different treatment of the concept is offered by Gerhard Hard in *Die "Landschaft" der Sprache und die "Landschaft" der Geographen. Semantische und forschungslogische Studien zu einigen zentralen Denkfiguren in der deutschen geographischen Literatur*, Colloquium Geographicum, vol. 11 (Bonn: In Kommission bei F. Dümmler, 1970), 71–79. The term *Stadtlandschaft* was coined by Siegfried Passarge in the mid-1920s.

21. The book appeared as *Die gegliederte und aufgelockerte Stadt* (Tübingen: Verlag Ernst Wasmuth, 1957).

22. See Dirk Schubert, "Theodor Fritsch und die völkische Version der Gartenstadt," *Stadtbauwelt* 73 (1982): 463–68. An extended version of this article was presented to the 2003 London meeting of the International Planning History Society, in which Schubert compares Fritsch and Ebenezer Howard. Fritsch died in 1933. There was an SA honor guard at his funeral, and his *Handbook on the Jewish Question* went through forty-three printings by 1943, sold hundreds of thousands of copies, and was admired by Hitler and Walter Darré, the Minister for Nutrition and Agriculture and prime spokesman for the ideology of *Blut und Boden*.

23. See Tilman A. Schenk and Ray Bromley, "Mass-Producing Traditional Small Cities: Gottfried Feder's Vision for a Greater Nazi Germany," *Journal of Planning History* 2, no. 2 (2003): 107–39.

24. The ordinance ("Richtlinien für die Planung und Gestaltung der Städte in den eingegliederten deutschen Ostgebieten," Allgemeine Anordnung no. 12/II, January 30, 1942) is reproduced in Durth and Gutschow, *Träume in Trümmer*, 45–50.

25. Bruno Wasser, *Die Neugestaltung des Ostens: Ostkolonisation und Raumplanung der National-sozialisten während der deutschen Besetzung in Polen 1939–1944* (PhD diss., Technische Hochschule Aachen, 1991); and Wasser, *Himmlers Raumplanung im Osten. Der Generalplan Ost in Polen 1940–1944* (Basel, Berlin, Boston: Birkhäuser, 1993). See also Uwe Mai, *"Rasse und Raum": Agrarpolitik, Sozial- und Raumplanung im NS-Staat* (Paderborn: Schöningh, 2002).

26. See Diefendorf, *In the Wake of War*, 170ff.; and Durth and Gutschow, *Träume in Trümmern*.

27. Niels Gutschow provided me with photocopies of the Wriezen protocols, taken partly from his father's papers, now housed in the Architektur Archiv in Hamburg, and Staatsarchiv Hamburg, Architekt Gutschow A44 (D38).

28. Gutschow, "Skizze Generalbebauungsplan 1944" (Manuskript Plassenburg), July 1944, presented at August meeting, 3.

29. Comment on Gutschow in ibid., 11.

30. Schelkes talk on "Grünflächenplanung," in ibid., 18.

31. Ibid., 22.

32. Ibid., 45, and Fritz Leonardt at the meeting of November 11–12, 1944, 32–33.

33. Protocol of September 16–17, 1944.

34. Martine Morel, "Reconstruire, dirent-ils. Discours et doctrines de l'urbanisme," in *Images, discours et enjeux de la reconstruction des villes françaises après 1945*, ed. Danièle Voldman, Cahiers du L'Institut d'histoire du temps present (Paris: Centre national de la recherche scientifique, 1987), 13–49, esp. 41.

35. Réme Baudoui, "Between Regionalism and Functionalism: French Reconstruction from 1940 to 1945," in Diefendorf, *Rebuilding Europe's Bombed Cities*, 43.

36. P. J. Larkham and Keith D. Lilley, *Planning the "City of Tomorrow": British Reconstruction Planning, 1939–1952—An Annotated Bibliography* (Pickering: Inch's Books, 2001); and Peter J. Larkham and Keith D. Lilley, "Plans, Planners, and City Images: Place Promotion and Civic Boosterism in British Reconstruction Planning," *Urban History* 30, no. 2 (2003): 183–205, esp. 189n30.

37. Larkham and Lilley, "Plans, Planners and City Images," 188.

38. Peter J. Larkham, "The Imagery of the UK Post-War Reconstruction Plans," Working Paper no. 88, Faculty of the Built Environment, School of Planning and Housing, University of Central England (Birmingham, 2004). See also Junichi Hasegawa, "The Rise and Fall of Radical Reconstruction in 1940s Britain," *Twentieth Century British History* 10, no. 2 (1999): 137–61; and Junichi Hasegawa, *Replanning the Blitzed City Centre: A Comparative Study of Bristol, Coventry, and Southampton, 1941–1950* (Buckingham, U.K., and Philadelphia: Open University Press, 1992).

39. Walter Gropius and Martin Wagner, "Cities' Renaissance," typescript in Houghton Library, Gropius Papers, BMS Ger 208 (56) [probably written by Wagner], 1.

40. Ibid., 5. Note the similarity between this terminology and that of Gutschow, who referred to "settlement cells."

41. Ibid., 9.

42. Ibid., 11.

43. Walter Gropius and Martin Wagner, "The New City Pattern for the People and by the People," epilogue to *The Problem of the Cities and the Towns, Conference on Urbanism, March 5–6, 1942*, 102.

44. Ibid., 115.

45. José Luis Sert, *Can Our Cities Survive? An ABC of Urban Problems, Their Analysis, Their Solutions: Based on the Proposals Formulated by the C.I.A.M., International Congresses for Modern Architecture, Congrès internationaux d'architecture moderne* (Cambridge, MA: Harvard University Press, 1942), 36–38, 66–68, 154. Eric Mumford, *The CIAM Discourse on Urbanism*, calls Sert's book a piece of propaganda, but he notes Sigried Giedion's defense of it as "the analysis of pathological conditions" (141). In 1944 CIAM members, including Gropius, Sert, and Richard Neutra, the head of the American CIAM chapter, met in New

York to try to identify for themselves a role in postwar reconstruction, but no consensus was reached at the meetings and no role proved forthcoming. See Mumford, *The CIAM Discourse*, 145–48.

46. W. G. Sebald, *On the Natural History of Destruction*, trans. Anthea Bell (New York: Random House, 2003), 4, 10, 6.

47. Jörg Friedrich, *Der Brand: Deutschland im Bombenkrieg, 1940–1945* (Berlin: Propyläen, 2002), 189, 519.

48. For just one interesting example, J. Van Rhijn, *Rotterdam, 1940–1946* (Delft: Delftsche Uitgevers Maatschappij, 1947), provides a photographic record of the destruction of that city by the Germans and concludes with two pages of drawings of a very modern "Rotterdam in the future" that features tower blocks, broad avenues bordered with green promenades, and people enjoying the waterfront.

49. Stadtarchiv Munich, Bauamt-Wiederaufbau, 1095, II, Section IX, nr. 1.

50. See Ishimaru Norioki, "Reconstructing Hiroshima and Preserving the Reconstructed City," in Hein, Diefendorf, and Yorifusa, eds., *Rebuilding Urban Japan*, 90.

51. Sert noted as early as 1942 the use of rubble in Rotterdam (*Can Our Cities Survive?*, 205). For Le Havre, see Joseph Louis Nasr, *"Reconstructing or Constructing Cities?" Stability and Change in Urban Form in Post–World War II France and Germany* (PhD diss., University of Pennsylvania, 1997), 71.

52. Hasegawa Junichi, "The Rebuilding of Osaka: A Reflection of the Structural Weaknesses in Japanese Planning," in Hein, Diefendorf, and Yorifusa, eds., *Rebuilding Urban Japan,* 74.

53. See Ishida Yorifusa, "Japanese Cities and Planning in the Reconstruction Period: 1945–55," in Hein, Diefendorf, and Yorifusa, eds., *Rebuilding Urban Japan;* Nick Tiratsoo, Junichi Hasegawa, Tony Mason, and Takao Matsumura, *Urban Reconstruction in Britain and Japan, 1945–1955: Dreams, Plans, and Realities* (Luton: University of Luton Press, 2002); and Emmanuel V. Marmaras, *Central London under Reconstruction: Policy and Planning, 1940–1959* (PhD diss., University of Leicester, 1992).

54. There were, of course, some significant restorable buildings that were either deliberately demolished or left in a demolished state. The demolition of the Stadtschloss in East Berlin was one; the refusal to rebuild Dresden's Frauenkirche was another.

55. Jacek Dominiczak, "Warsaw and Gdansk as Two Distinctive Approaches to Post–Second World War Reconstruction: Urban Design and the Problem of Method," in *Urban Triumph or Urban Disaster? Dilemmas of Contemporary Post-War Reconstruction,* ed. Sultan Barakat, Jon Calame, and Esther Ruth Charlesworth (York: Post-War Reconstruction and Development Unit, University of York, 1998).

CONTRIBUTORS

GREG BANKOFF, professor of modern history at the University of Hull, is a social and environmental historian of Southeast Asia and the Pacific. In particular, he writes on environmental-society interactions with respect to natural hazards, resources, human-animal relations, and issues of social equity and labor. Among his publications are *Crime, Society, and the State in the Nineteenth-Century Philippines* (1996) and *Cultures of Disaster: Society and Natural Hazard in the Philippines* (2003). He is also coeditor, with Georg Frerks and Dorothea Hilhorst, of *Mapping Vulnerability: Disasters, Development, and People* (2004). His most recent works include a volume coedited with Peter Boomgaard, *A History of Natural Resources in Asia: The Wealth of Nature* (2007), and a book coauthored with Sandra Swart, *Breeds of Empire: The "Invention" of the Horse in Maritime Southeast Asia and Southern Africa, 1500–1950* (2007).

LISA M. BRADY is associate professor of history at Boise State University, where she teaches courses on North American and global environmental history. Her research focuses on the environmental implications of war and peace in a variety of contexts. She is author of "The Wilderness of War: Nature and Strategy in the American Civil War" (*Environmental History*) and "Life in the DMZ: Turning a Diplomatic Failure into an Environmental Success" (*Diplomatic History*). Her book on the American Civil War is forthcoming, and she is currently working on an environmental history of conflict in twentieth-century Korea.

DOROTHEE BRANTZ is assistant professor of history at the Center for Metropolitan Studies at the Technische Universität Berlin. Her research areas include urban and environmental history as well as the history of warfare. She has edited two volumes on the history of human-animal relations, and her essays have appeared in *Central European History,* the *Bulletin of the German Historical Institute, Food and History,* and several edited volumes. Her current project focuses on the role of the environment in total warfare during the twentieth century.

CHARLES E. CLOSMANN is assistant professor of history at the University of North Florida, where he teaches courses on the history of modern Europe, Nazi Germany, and the global environment. His research interests include the environmental history of Germany, urban environmental history, and the history of fisheries. He has published articles in the *Journal of Urban History,* the *Hamburg Wirtschaftschronik,* and the *Bulletin of the German Historical Institute,* and is currently writing an environmental history of the city of Hamburg.

JEFFRY M. DIEFENDORF is the Pamela Shulman Professor of European and Holocaust Studies at the University of New Hampshire. He is the author of *In the Wake of War: The Reconstruction of German Cities after World War II* (1993), editor of *Rebuilding Europe's Bombed Cities* (1990), and coeditor of *Rebuilding Urban Japan after 1945* (2003) and *City, Country, Empire: Landscapes in Environmental History* (2005). Most recently he published "Reconciling Competing Pasts in Postwar Cologne" in *Beyond Berlin,* edited by Paul Jaskot and Gavriel Rosenfeld (2007). In addition to an ongoing study of mid-twentieth-century Cologne, Basel, and Boston, he is working on the art of Holocaust survivor Samuel Bak and a comparison of the rebuilding of bombed cities and the rebuilding of post-Katrina New Orleans.

MARCUS HALL is assistant professor of history at the University of Utah and associate researcher of environmental sciences at the University of Zurich. He is also the author of *Earth Repair: A Transatlantic History of Environmental Restoration* (2005), which received the Downing Book Award from the Society of Architectural Historians. In addition to warfare, Hall is pursuing various historical questions related to malaria, exotic species, salvage archaeology, and environmental restoration.

J. R. MCNEILL, professor of history and University Professor at Georgetown University, is the author of *Epidemics and Geopolitics in the American Tropics, 1640–1920* (2009), *Something New under the Sun: An Environmental History of the 20th-Century World* (2000), and *The Mountains of the Mediterranean World: An Environmental History* (1992), along with several other books and numerous articles in the fields of environmental history and international history. He is also the coauthor, with William H. McNeill, of *The Human Web: A Bird's-Eye View of Human History* (2003).

DAVID S. PAINTER teaches international history at Georgetown University, where he has a joint appointment with the Department of History and the School of Foreign Service. His research focuses on the international oil industry, the Cold War, and U.S. relations with the third world. He is the author of *Oil and the American Century: The Political Economy of U.S. Foreign Oil Policy, 1941–1954* (1986) and *The Cold War: An International History* (1999) as well as the coeditor of several essay collections. In the spring of 2008 he was a Senior

Visiting Fellow at the Nobel Institute in Oslo, Norway, working on a forthcoming study of oil and global power.

CHRIS PEARSON is a research associate in historical studies at the University of Bristol, where he is completing an environmental history of military bases and training grounds in contemporary France. His main research interests are the environmental history of Europe (especially France), war and the environment, animal studies, and human-nonhuman relations. He has published essays in *Environmental History* and *Revue forestière française* and is the author of *Scarred Landscapes: War and Nature in Vichy France* (2008).

FRANK UEKÖTTER is a Dilthey Fellow with the Research Institute of the Deutsches Museum in Munich, Germany. His research interests include the history of environmental policies in Germany and the United States, the history of environmental movements, and the history of modern agriculture. His books include *The Green and the Brown: A History of Conservation in Nazi Germany* (2006) and *The Age of Smoke: Environmental Policy in Germany and the United States, 1880–1970* (2009).

ROBERT WILSON is an assistant professor of geography at the Maxwell School of Citizenship and Public Affairs at Syracuse University, where he teaches courses in environmental history and historical geography. He is the author of *Seeking Refuge: An Environmental History of the Pacific Flyway* (in press) and is currently working on an environmental history of Japanese American internment camps during World War II.

INDEX

Entries in *italics* indicate illustrations.

Addis, Michael, 153
Adventure in Sardinia, 122
aerial attacks:
 environmental damage caused by, 27
 loss of life caused by, 185
 as modernization opportunity, 6, 176, 181, 182, 184
 property damage caused by, 41, 172–73, 184–85
aerial herding, 6, 139–40
Agent Orange, 2, 22
Agro Pontino, 117
Ahern, George, 34
Aisne River, 72
Albay, 37
Algiers, 121, 182
Allied air raids, 6, 177, 180
Allied Commission, 117
Allied Forces, 20–21, 94, 112, 113, 116–17, 118, 119, 121, 126, 156
Allied soldiers, 18, 176
Alps, 84, 150, 154–55, 158
Alps at War, The, 160
Alsace, 178–79
American Revolution, 2, 15
American Society for Environmental History (ASEH), 2
"American system of manufacture," 14
Amerindians, 10–11, 13. *See also* Native Americans
Amsterdam, 23
Andrews, Eliza, 62–63

Anhalt region, 97
animals, impact of war on, 81–82, 85–86n13
Annales School, 8, 71
Anopheles, 121, 124, 127
Anopheles gambiae, 121
Anopheles labranchiae, 117, 124
antimalaria, experimental drugs, 5, 119
antimalaria measures, 2–3, 5, 112–29
antimalaria measures, hindered by war, 116
Appalachian Mountains, 13
Arbeitsstab Wiederaufbauplanung bombenzerstörter Städte (reconstruction-planning task force), 180
Architectural Forum, 175
Argentina, 13
Argonnaute, L', 78
Argonne, Forest of the, 72
"armory system," 14
Army of Northern Virginia, 61
Asia, 20, 21, 22. *See also* East Asia, Southeast Asia
Association of German Architects (Bund Deutsche Architekten), 175
Asylum of Santa Maria della Pietá, 119
Atlanta, 50, 52, 55
atomic bombs, 19, 185
atomic weapons, 20, 25
atomic weapons testing, environmental effects of, 25, 26
Auschwitz-Birkenau extermination camp, 99, 153, 160
Australia, 13

Aux Armes!, 159–60
Averasboro, 60, 61
Axis Europe, 184–85

Baden, 92–93, 94
Badisches Ministerium des Kultus und Unterrichts als Höhere Naturschutzbehörde (Baden minister of education and cultural affairs), 93
Baltic Sea, 72
Balzarro, Anna, 160
Bankoff, Greg, 4, 32–48, 193
Barrière, Philippe, 157
Bataan, 42
Bates, Marston, 124, 126, 128
Battaglie di Pace (Battles of Peace), 122
Bauhaus movement, 183
Bavaria, 97
Beaufort, 57, 58, 60
Belagerungszustandsgesetz (state-of-siege law), 98
Belgian cities, destruction of, 173, 180
Belgian forests, use of for trench building, 74
Belgium, 71
Bellinchen, 100
Benterberg, 178
Bentonville, 60
Berlage, H. P., 173, 180
Berlin, 6, 92, 93, 100, 173, 175, 176, 180, 185, 186, 187, 188
B. F. Goodrich Chemical Company, 143
Bielefeld, 99
Bielitz, 99
bird management, 134, 141
Birori, 125
Bitterfeld, 99–100
Bloch, Marc, 71, 76–77, 79
Bochophage, Le, 79
Bochum, 185
Bongsu, Rajah, 35
Bonifica Integrale, 114
Boston, 188
Bourn, Warren S., 143
Boy Scouts, 71
Brady, Lisa M., 4, 49–67, 193
Brandstätten, 177

Brantz, Dorothee, 4, 68–91, 193
Brazil, 13, 121
Bremen, 96, 97, 185
Britain, 11, 17,21, 25, 72, 126, 171, 175, 182, 186
British, 11, 27, 77, 79, 80, 82, 85–86, 113
British Howitzer, 74
British Royal Navy, 15
Bruller, Jean, 162
Brussels, 175, 183, 188
Bund Deutsche Architekten (Association of German Architects), 175
Bundesrat, 95
bunkers, as protection from aerial bombings, 177
Bureau of Forestry (Philippines) (Inspección General de Montes), 34

Cagliari, 127
Calabria, 118
California, 5, 11, 25, 133–39, 142, 145
California Department of Fish and Game, 135, 137
Camarines, 37
Can Our Cities Survive? An ABC of Urban Problems, Their Analysis, Their Solutions: Based on the Proposals Formulated by the C.I.A.M., 183
Caracoas (Philippine warship), 35
Carolinas, 49, 50, 51, 56, 57–63. See also Sherman's 1864–65 Campaigns
Caribbean, 16–17
Carson, Rachel, 145
Cassino, 121
Castel Volturno, 112, 115, 121
Caucasus region, 101
Cavite, 37
Central America, 16–17
Central European Pipeline System (CEPS), 23
Chapelle-en-Vercors, La-, 154, 155
Charlesworth, Andrew, 153
Charter of Athens, 172, 175, 178, 182
chemical weapons, use of, 18–19, 22, 80
chemical weapons industry, 18–19
Chemnitz, 95

Index 199

chevauchée (massive foraging raid), 50, 52, 61, 62, 63. *See also* foraging
Chicago, 14, 140
Chicago Edison Company, 98
China, 16, 19, 105
cholera, 172
chlorine gas, 18, 80–81, 142
Christaller, Walter, 179–80
Christians, 36, 38
Christian Democrats (Italy), 126
CIAM (Congrès Internationaux d'Architecture Moderne [International Congress of Modern Architecture]) 172, 175–76, 178, 182, 183
cities, as threats to life and civilization, 172
cities, civil defense of, 176;
    evacuation and bunker plans, 177
    firebreaks, as preventative measure, 176–77
    influenced by modernist design, 178
    protection from aerial bombings, 176, 177–78, 181
cities, duty of, to protect health, 174
cities, green spaces in, 172–75, 177–78, 179, 180, 181
cities, historical origins of, 172
cities, post World War I hardships in, 173
cities, rebuilding as opportunity for modernization, 6, 171, 173, 176, 177–78, 180
cities, World War II damage to, 173, 176, 177, 184–85
"Cities' Renaissance," 183
city beautiful, as urban planning model, 171
City of the Future: Garden City, The (*Die Stadt der Zukunft* [*Gartenstadt*]), 179
cityscapes, 171, 176, 184, 187
Civilian Conservation Corps (CCC), 133, 135, 137, 141–42
Clark Air Base, 22
Closmann, Charles Edwin, 1–9, 194
Col de La Chau memorial, 160–62, 164, 165
Cold War, 12–13, 21, 23–24, 25, 26, 126, 156
collateral damage, defined, 134
collateral productivity, defined, 134
Cologne, 96, 177, 185, 186, 187
Colonia Militar of Tumauini (fort), 39, *40*

Colorado, 25, 26
Columbia (South Carolina), 56, 57, 58, 59, 60
Colusa, 138, 139
Colusa National Wildlife Refuge, 139
Coluzzi, Alberto, 114, *115*, 118
Combés, Francisco, 35
Comité d'aide et de reconstruction de Vercors, 155
Communists, 126, 187
Confederacy, 49–51, 52, 54, 56, 58, 62, 63;
    agricultural foundations of, 4, 49, 54, 55
    ecological foundations of, 51, 55
Confederate Army, 4, 51, 54, 60
Confederate populace, 62
Congrès Internationaux d'Architecture Moderne (International Congress of Modern Architecture [CIAM]), 172, 175–76, 178, 182, 183
Conservation in Action, 145
conservation programs:
    during Nazi Germany, 5, 92–95
    pre-WWII in U.S., 134–37
    impact of WWII on, in U.S., 133–34, 135–40
Coosawatchie River, 52, 58
Corbusier, Le, 172, 175, 182
Cordilleras (region of Luzon), 33, 28, 29, 43
Corsica, 124
Cosgrove, Denis, 69
Couturier, Marcel, 154
Coventry, 6, 182, 186
Cox, Maj. Gen. Jacob, 53, 54, 58, 59, 61
crematoriums, 99, 183
Crete mosquito eradication initiative, 121–22
Crosby, Alfred, 13
Cuba, 16, 17
Cyprus mosquito eradication initiative, 121–22

Dalloz, Pierre, 156
Darfur, 1
Darré, Walter, 179
DDD, 144

DDT (dichloro-diphenyl-trichloroethane):
    known dangers of, 143
    discovery of, 19
    efficacy of, 118, 120, 124
    environmental damage of, 19–20, 126–29
    impact on fish and wildlife, 2–3, 19–20, 126–29, 144–45
    use by FWS in U.S. refuges, 143–44
    use against malaria, 19, 112–29
    use in Sardinia, 5, 112, 118–22, 124–28
    use against typhus, 19, 112, 119, 120, 121
Defense Energy Support Center, 24
Defense Science Board, 23
defoliants, use of, 8, 22, 27. *See also* Agent Orange
deforestation, in Philippines, 32–35, 36–38, 43;
    as a result of fort construction, 38–40
    as a result of Japanese occupation, 40–42
    as a result of shipbuilding, 36–38
    consequences of, on microecosytems, 43
denazification, 181
dichloro-diphenyl-trichloroethane. *See* DDT
Diefendorf, Jeffry M., 6, 171–92, 194
dieldrin, 144
Dortmund, 185
Dresden, 19, 171, 185, 187
Drôme, 155
Duisburg, 96
Dutch, 33, 36, 38, 43
Dutch East India Company (Vereenigde Oostindische Compagnie [VOC]), 36
dysentery, 71, 112

East Asia, 16
East Berlin, 187
"ecological imperialism," 13
ecosystems, 3, 4, 6, 19, 21, 28, 112, 127, 128, 129
Egestorf, 100
Egypt, 121, 185
Eighty Years' War, 36
Ente Regionale per la Lotta Anti-Anofelica di Sardegna (ERLAAS), 5, 120–26

entomology, 11, 124, 126, 128
entrenchment landscapes, 72–75
environment, definition of, 1, 7n3, 68–69
environment, relationship with weaponry, 6, 42, 81
"environment of war," 69
environmental forces, role in war, 69
environmental history, definition of, 7n3, 8n8
environmental history, link with urban planning, 171
environmental history, neglect of warfare, 2, 70
environmental impact of war, indirect effects, 24–25, 132, 135–40, 145–46
environmental protection, during wartime, 5–6, 100–103, 132–33, 135–40, 145
environmental remediation, 1, 27
epidemics, 80, 119, 120, 171, 172
Erfurt, 95, 96
ERLAAS (Ente Regionale per la Lotta Anti-Anofelica di Sardegna), 5, 120–26
erosion of species (genetic), 4, 33–34, 40, 43
Ewing, Philemon, 55

Farmer, Sarah, 152
Fascists, 120
favism, 128
febbre, le (the fevers), 114
Feder, Gottfried, 179
Federal Agency for Water and Air Quality (Reichsanstalt für Wasserund Luftgüte), 99, 100, 102, 104, 105
Filou, Le, 81
fire, as military tool, 51
Five Power Naval Treaty, 16
Flanders, 72, 79
Foeddu, Giuseppe, 125
Foote, Kenneth, 151
foraging, 50, 52–53, 55, 57, 60, 62;
    as war right, 58–59
Ford, Henry, 14
forest death, 96, 102
Formosa (Taiwan), 36
fort construction, 4, 32, 33, 38–40, 43
Foxworthy, Fred, 39

Index

France, 6, 11, 14, 23, 25, 82
France (Louis XIV), 10
Frankfurt, 23
Free French, 156
Freiburg University, 105
French Alps, 154–55
French Forestry Service, 82
French Resistance, 150–60, 164–66
French Resistance memorials, 150–53, 155, 157, 158, 162–66
Freudenstadt, 187
Friedrich, Jörg, 177, 184
Fritsch, Theodor, 179, 190n22
Frontschwein, 73

game animals, 132–33
garden city, 171, 174, 179, 188, 189n9
garden suburbs, 171, 175, 188
gases, poisonous, use in World War I, 18–19, 74–75, 80–81, 90n85, 142
Gdansk, 187
Gehrdenerberg, 178
Georgia, 4, 49–63
Georgia State Railroad, 55
Generalinspektor für Wasser und Energie (Inspector General for Water and Energy), 93
Generalinspektors für das deutsche Straßenwesen (Inspector General for German Highways), 93
Generalplan Ost, 179
Geneva Protocol of 1925, 19
German Academy for City, Reich, and Regional Planning, 178, 180
German Bureau for Nature Protection (Reichsstelle für Naturschutz), 100
German Department of the Interior (Reichsministerium des Innern), 102, 104
German Landscape Advocate (Reichslandschaftsanwalt), 93
German Ministry of Trade and Commerce (Reichswirtschaftsministerium), 98, 99, 102
German rearmament, 98, 99
German society of engineers (VDI-Fachausschuß für Staubtechnik), 100, 101

German trading regulations (Reichsgewerbeordnung), 97
Germany, 18, 71, 102, 178, 183;
    bureaucratic aspects of environmental decisions in, 94–101, 102–6
    conservation programs in, 5, 92–95, 103, 104
    environmental policy of, during World War II, 94, 99–103, 104–5
    environmental policy of, pre-WWII, 95–99
    rebuilding of cities in, 6, 184–86
    pollution controls in, 96–99, 100, 101, 102–4
    urban planning in, 171, 178, 181, 182
    World War I, 72, 92
    World War II damage to, 19, 105, 174, 184
    Wutach Gorge project conflict in, 95, 104–6
Gibson, Donald, 182, 186
Gilioli memorial, 151
Göderitz, Johannes, 178, 181
Goebbels, Joseph, 92
Goldsboro, 57, 60–61
Gordon, 62
Göring, Hermann, 103
Gran Cordillera Central mountains, 38
Grant, Ulysses S., 50, 56–57, 58, 61, 64n5
Graubner, Gerhard, 177, 178
Great Lakes, 14
Greece, 124
green spaces, 172–75, 177–78, 179, 180, 181
"greening" cities, 173
Grenoble, 162
Grenoble Museum of Resistance and Deportation, 160
Gresse-en-Vercors monument, 151, 152
Griffith, Richard E., 144
Gropius, Walter, 183
groundwater contamination, 22, 26, 27
Gruerie, La, 77
Guantánamo Bay, 16
Guidelines for the Planning and Design of Cities in the Annexed German Territories in the East, 179

Gulf War (1991), 8n9, 24, 26, 27
Gutach River, 92
Gutschow, Konstanty, 178, 180

Habeck, Mary, 76
Hall, Marcus, 5, 112–31, 194
Halleck, Henry Wagner, 52, 55, 67n73
Hamburg, 6, 19, 100, 177, 180, 185, 186
Hampton, Lt. Gen. Wade, 58–59
Hanford Engineering Works, 26
Hannover, 178
Hardee, Gen. William J., 54
hardwoods, 14, 33, 35, 38, 39, 41, 43, 44n8, 133
Harney Basin, 142
Harpers Ferry, 14
Haslach River, 92
Havre, Le, 182, 186
Haushalt der Natur (natural economy), 181
Hawai'i, 16
Hay-Bunau Varilla Treaty, 17
Hayasi, Maj. Gen. Yoshihide, 41
hazardous waste, 1, 21, 22
Hegel, 103
Hegemann, Werner, 172
Heidelberg, 23, 117
Heldenhaine, 84
Hellerau, 171, 189n9
herbicides, 27, 28, 142–43, 145;
    2,4-D, 142, 143
    VL-600, 142.
    See also DDT
Hercules, 59
Herdecke power plant, 101
Hermann Göring steel works, 181
Hideaki, Ishikawa, 187
Himmler, Heinrich, 179, 180
Hippocratic Oath, 119
Hiroshima, 185, 187
History of the Peloponnesian War, 1
Hitchcock, Henry, 53–55, 57, 58, 61, 64n15
Hitler, Adolf, 172
Hoffmann, Hubert, 178
Holden, Charles, 186–87
Holford, William, 186–87
Holocaust memorials, 151, 167n12

Homer-Dixon, Thomas, 43
Horn, E. E., 139
House, Col. Edward, 17
Howard, Ebenezer, 171, 179
Humphreys, Margaret, 124–25
Hüppauf, Bernd, 78
hydrogen bomb, 20, 25

Ickes, Sec. of Interior Harold, 133
"The Idea of War as the Basis of City Form and City Planning," 177
Illinois State Council of Defense, 98–99
Imperial Army (Japan), 41
Imperial German government, 98
Imperial Valley, 139, 142
incendiary bombs, mass production of, 19
Indian Ocean, 22
influenza, from trench warfare, 80
Inspección General de Montes (Philippines forestry department), 34
Inspector General for German Highways, 93
Insull, Samuel, 98–99
Iran-Iraq War (1980s), 27
Iraq, 27
Iraqi forces, 27
irrigation districts (U.S.), 141–42
Isabela province, 39
Istituto Superiore di Sanità, 114, *116*
Italy, 5, 112–31
Iwo Jima, 26

Jansen, Marcel, 159
Jansen, Paul, 164–65
Japan, 4, 15, 19, 21, 25, 33, 36, 40–43, 137, 176, 184, 186, 188
Japanese American internment camps, 140–41
Japanese-led deforestation in Philippines, 40–42
Java, 33, 36
Johnston, Gen. Joe, 49, 60–61
Jolo, 35–36
Jordana y Morera, Ramón, 37
Joseph, Gilbert, 160
Jünger, Ernst, 68, 76–77, 81
Jurisch, Konrad Wilhelm, 96

Index

Kattowitz, 102
Kerr, John, 125
Kiel, 174–75
Klamath Basin, 134–35, 137, 139–40, 142, 145
Klamath Irrigation Project, 135, 140, 141
Klamath River, 134
Klose, Hans, 100, 101, 105
Königsberg (Kaliningrad), 187
Königshütte, 102
Korean War (1950–53), 20, 22, 25
Kosovo, 2, 24
Kreisleitung (NSDAP's district organization), 99
Kriegsbegeisterung (enthusiasm for war), 71
Kries, Wilhelm, 175
Krull, Christian, 78
Krupp industrial empire, 96
Kuwait, 27

Laboratory of Parasitology, 118
Larkham, Peter, 182
LeConte, Emma, 56
LeConte, Joseph, 60
landmines, 27
landscape, definition of, 68–69
landscape, changing due to trench warfare, 75, 84n6
Landshaft, 68
Landschaftsanwälte (landscape advocates), 105
Latin America, 18, 22
Laurel, Jose, 42
Lawson, Alexander, 62
Lea Act, 139
Lee, Gen. Robert E., 55, 57, 61
Leipzig, 99
Leningrad, 187
Leonhardt, Fritz, 181
Letchworth, 171
Leuven, 173, 180
Leyte, 37
lice, 79–80, 119, 143. See also louse, typhus
Lilley, Keith, 182
Lincoln, Frederick, 135
Lincoln, President Abraham, 54
Little Ogeechee River, 54

Loddo, Mariantonia, 125
Logan, John, 124
logging, 34, 40, 41–42, 74, 133
Loire Valley, 182
London, 175, 182, 186
London Regiment, 80
Loos, Battle of, 77
louse, 19, 79, 80, 119, 120, 121. See also lice, typhus
Lower Klamath Lake, 135
Lübeck, 180, 185
Lucherberg, 96
Lüdenscheid, 101
Luire cave, 164–65
Lumber River, 62
Luzon, 33, 34, 37, 38, 41, 42

MacArthur, Gen. Douglas, 113
MacArthur's Pacific campaign, 113
MacDonald, George, 119
Maccarese, as location of DDT studies, 118
Maccarese (Tiber delta), 117, 118
Maccarese Reclamation Company, 118
machine-age warfare, 4, 6, 10, 75, 81
Maguindanaos, 35
Mahan, Alfred Thayer, 15
Mairet, Louis, 73
Maison Carrée Prison, 121
Makiling National Park, 42
mal-aria (bad air), 114. See also malaria
malaria, 2–3, 17, 19;
    combating, in Sardinia, 5, 112–29
    manifestations of, 114
    mortality from, 115
    as cause of social unrest, 114, 115–16
    susceptibility of soldiers to, 116
    symptoms of, 114
    increased by warfare, 112, 115
malaria, as weapon of warfare, 117–18
Malaria Control Demonstration Unit, 121
malaria therapy (testing), 119
malariologists, 5, 114, 117, 119, 125
malariology, 118, 127
Maldonado, Juan, 36
Malheur River, 134
Malleval, 155

Manila, 16, 34, 36, 37, 39
Manzanar War Relocation Center, 141
maquis, 153, 154, 155, 156–60, 164, 165
maquisards (resistance fighters). *See* maquis
Marc, Franz, 75–76, 77, 79
Marchon, Albert, 157
Marinduque, 37
Marne, Battle of the, 71–72
Marsh Festival (1955), 128
Martini, Erich, 117;
    role in fostering malaria epidemic, 118
Masbate, 37
McNeill, J. R., 3, 10–31, 194
medical testing, malaria, 118–19
Mediterranean, 23, 126
memorials, 150–53, 155, 157–58, 162, 163–64, 165, 166
Mestre, Philippe, 166
Mexico, 11, 34
Meuse River, 96
Meyer, Konrad, 179
Middle East, 18, 21
Midway Island, 16
Midwest (America), 132
Migge, Leberecht ("Spartacus in Green"), 173, 174–75
migratory birds, impact of war upon, 5–6, 134, 137, 138, 141, 145, 146
military industrial complex, 2
military history, neglect of environmental aspects of war, 2, 70
Milledgeville, 55, 62–63
Mindanao, 34, 38
Minister für Wissenschaft, Kunst, und Volksbildung, (Prussian Minister of Science), 100
Ministerio de Ultramar, 40
Missiroli, Alberto, 114, 115, *116*, 117–19, 121, 124, 125;
    collaboration with German authorities to foster malaria epidemic, 117–18
    role in testing DDT, 118
Mississippi River, 14, 64n5
Missouri River, 14
Modern Architecture Research group (MARS), 182
modernist functionalism, 171–72, 173, 178, 181–82, 183, 187, 188
Moluccas, 36
Mommsen, Wolfgang, 73
monumentalism, 171, 178, 181, 187
Morga, Antonio, 38
Moron (airfield), 23
Moro raiders, 33, 36, 38, 43
Moscow, 175, 179
mosquitoes, 2–3, 5, 17, 19, 78, 112, 115–29. *See also* malaria
Mosse, George, 84
motorization of warfare, 17–18
Moucherotte, Le, 158
"M.3," 119
Mughal Empire, 3, 10
Muir, John, 129
Müller, Paul, 125
Mumford, Lewis, 128
Munich, 95, 185, 186
Mussolini, 113–14, 172

Nagoya, 176–77, 187
Naples, 112, 119, 120–21
Napolean, 113
Nash, Peter, 154
National Conservation Law of 1935, 92, 103
National Socialism, 103
National System of Interstate and Defense Highways, 25
National Wildlife Refuges in California, 136
Native Americans, 132. *See also* Amerindians
Natural Enemy, Natural Ally: Toward an Environmental History of War, 3
natural resources, 1, 2, 3, 6, 7, 43, 69, 94, 177;
    demand for during wartime, 5–6, 74, 132–33, 145
    use of for trench building, 74
Nazi destruction of Tiber delta's pumps, 117
Nazi Party (National Socialist German

Workers' Party [NSDAP]), 98, 99, 103–5, 120, 179;
   conservation policies of, 5, 93–94, 95
   urban plans of, 175–76
Negros (Philippines), 34
Nemico Fulminato, il (the Annihilated Enemy), 122
Netherlands, 36, 187
Netherlands East Indies, 18
neue Stadt, Die (*The New City*), 179
Neufert, Ernst, 180
Neumann, Franz, 103
Nevada, 25–26
New Deal, 133, 134
New England, as military enclave, 25
"new imperialism," 15
new military history, 2, 3, 7, 70
New Orleans, 14
New Zealand, 13
Nichols, George Ward, 52, 55, 59
Niemeyer, Reinhold, 180
Nile River, 121
Noguères, Henri, 157
"no-man's-land," 74, 75, 83–84
North Africa, 26
North Carolina, 51, 57, 60–61
nuclear arsenal, 20
nuclear deterrent, 21
nuclear superiority, 20
nuclear waste, 26, 28
nuclear weapons, 10, 23, 25
Nuremberg, 185

Oakey, Daniel, 56, 57, 61
Oberkrome, Willi, 94
Ohio River, 14
oil consumption by military, 18, 23–24
Okinawa, 21, 26
Olympic Peninsula, 133
Oradour-sur-Glane, 152–53
ordnances, depleted uranium, 27
Oregon, 11, 134, 137, 142, 147n8
Orr, Lord Boyd, 126
Ortsgruppe als Siedlungszelle, 178
Ortueri, 125
Osaka, 186

Oton, 37
Ottoman Empire, 10
Owens Valley, 141

Pacific Coast, 140
Pacific Flyway, 134, 135, 137
Pacific islands, 26
Pacific Northwest, 25, 133
Pacific Ocean, 16, 17
Pacific region, 17, 21
Pacific theater, 19, 26, 113;
   use of DDT in, 143
Painter, David S., 3, 10–31, 194
paludismo (swamp disease), 114
Paludrine, 119
Panama, 17, 22
Panama canal treaties, 22
parasites, 112, 119
Parc naturel régional du Vercors (PNRV), 150, 155, 158, 160, 164, 166n4
Paris, 71
Paris Green, 118–19, 121, 122
Parliament (Germany), 95, 97, 98
Pas de l'Aiguelle memorial, 165
Pasteur Institute, 121
Patuxent Wildlife Research Center, 143
Pearl Harbor, 16;
   attack on, 19, 41
Pearl River, 16
Pearson, Chris, 6, 150–70, 194–95
Pearson, Michael, 151
Pericolo Sovrastante, il (the Overwhelming Danger), 122
Perret, August, 182
Persian Gulf, 21, 22, 27
Philippines:
   annexation by U.S., 16
   deforestation of, 4, 32–43
   fort construction in, 38–40
   genetic erosion in, 4, 33–34
   pre-Hispanic period, 34–36
   Japanese occupation of, 40–42
   commercial logging in, 34
   scarcity of lumber in, 34
   shipbuilding in, 36–38
   impact of war on environment, 33–43

Philippine Lumber Control Union, 41
Pickelhauben (spiked helmet), 75
Pinatubo, Mt., eruption of, 22
Pionniers du Vercors (Vercors veterans association), 154, 155, 157, 165
Pionnier du Vercors, Le, 155
Plan Montagnards, 156
plantation agriculture, 49
Plasomochine, 119
plutonium, 25, 26
PNRV (Parc naturel régional du Vercors), 150, 155, 158, 160, 164, 166n4
Pocotaligo, 58, 61
poilu, 73
Poland, 178–79, 184
Pont-en-Royans, 155
poison gas, use of, 18–19, 74–75, 80–81, 90n85, 142
Polizei Auschwitz, 99
pollution control, in Germany, 96–97; pre–World War II, 96–98 during World War II, 99, 100, 101, 102–4
Poperinge, 81
Porta Westfalica, 100
Portotorres, 118–19
Posada, 118–19
Prefect of the Isère, 154
Provence, 156
Prussia, 3, 98, 99
Prussian Agency for the Preservation of Natural Monuments, 99–100
Prussian Institute for Water, Soil, and Air Hygiene, 104–5
Prussian Minister of Science (Minister für Wissenschaft, Kunst, und Volksbildung), 100
Prussian Ministry of Education, 96
Prussian Ministry of War, 98
Puerto Rico, 16, 22
pyrethrum, 116–17

quinine, 116, 119, 125, 126, 127

Raffarin, French Prime Minister Jean-Pierre, 150
Rainer, Roland, 178
Raleigh, 57, 61
rat infestations in trenches, 79–80
Raumordnung (regional planning), 180–81
reconstruction, in postwar Germany, 6, 184–86
Red Army, 100
refugees, 1, 116, 117, 153
refuges, for wildlife conservation, 5, 6, 133–146
Reich, 94, 98, 99, 100, 178, 181
Reich Office for Regional Planning, 180
Reichow, Hans Bernhard, 178
Reichsanstalt für Wasserund Luftgüte (Federal Agency for Water and Air Quality), 99
Reich's Forest Service (Reichsforstmeister), 93
Reichsgesundheitsführer (Secretary of State for Public Health), 105
Reichsgewerbeordnung (German trading regulations), 97
Reichsministerium des Innern (Department of the Interior), 102
Reichsstelle für Naturschutz (German Bureau for Nature Protection), 100
Reichstag (Parliament), 95, 97, 98
Reichswirtschaftsministerium (German Ministry of Trade and Commerce), 98, 99
remediation costs, 1, 27
Remscheid, 185
Rester-Résister garden of memory, 162–64
Rhein-Main Air Base, 23
Richmond, 57
Rimpl, Herbert, 181
Ríos Coronel, Hernando de los, 38
Rizal, José, 38
Rochelle–La Pallice, La, 182
Rockefeller Foundation, 5, 114–15, 117, 120–22, 124, 126, 127, 129
Rocky Flats Arsenal, 26
Rohrscheidt, Kurt von, 97
Roman Campagna, 112, 114
Rome, 114, 117, 118, 119
Roosevelt, President Franklin D., 19, 133, 135
Ross Institute, 119
Rota, 23

Index

Rothenburg ob der Tauber, 187
Rotterdam, 186, 188, 192n48
rubble, due to war, 82, 116, 127, 176, 177, 185
rubble, postwar removal of, 185–86
Ruhr Basin, 101
Ruhr Valley, 96
Russell, Edmund, 2, 3, 122
Russell, Paul, 117, 120
Russia, 105, 113

Sacramento National Wildlife Refuge, 135, 137, 138, 139, 142, 144
Sacramento Valley, 137, 138, 142, 143, 144
Saint-Dié, 182
Saint-Martin-en-Vercors, 154–55
St. Mihiel, 96
Saint-Nizier-du-Moucherotte, 154, 157, 159
Saipan, 26
Saliers, 162
Salkahatchie River, 58, 59
Salyer, J. Clark, II, 140, 141
Samar, 37
sandstorms, as caused by tank warfare, 26
San José Island ("Test-Tube Island"), 22
"The Sanitary Greenery of the City," 173
Sardinia, 5, *113*, 114–15, 118–29
Sardinia Project, 5, 115, 120–29;
    damage to ecosystem, 126, 128
    amount of DDT sprayed, 126
    methods compared, 125
    politics of, 126
    success of, 124
    perceived success of, 125
Sardinian Project, The, 122
sarin, 22
Sauer, Carl, 128
Savannah, 50, 55, 57–58, 62;
    topography of, 54
Schelkes, Willi, 181
Schicksalsgemeinschaft (community of fate), 84
Schluchseewerk project, 92–93, 94
Schluchseewerk power company, 92–93
Schrebergärten (truck gardens), 1le74
Schurhammer, Hermann, 92–93, 94, 104–5
Schutzstaffel (SS), 99

Scott, James C., 3
Sebald, W.G., 184
Secretary of State for Public Health (Reichsgesundheitsführer), 105
Seeing like a State: How Certain Schemes to Improve the Human Condition Have Failed, 3
Seifert, Alwin, 93
Sequoia National Park, 133
Serbia-Kosovo bombing campaigns (1999), 24
Sert, José Luis, 183, 191–92n45
Sevastopol, 187
Sgard, Anne, 158
Shell Petroleum Company, 122
Sherman, Gen. William T., 4, 49–63
Sherman's 1864–65 Campaigns:
    Carolina Campaigns, 57–62, 63
    "March to the Sea," 4, 50, 51–56, 62–63
    purpose of, 49, 50–52, 53, 58–59, 62, 63, 64n5
    route of, 52, 54, 56,
    success of , 54–55, 61–63
shipbuilding, 15–16, 20, 33, 36–40, 43
Shoshone Indian lands (Nevada), 25
Sicily, 113, 117, 121
Silent Spring, 145
Silva, Juan de, 38
Silvestre, Paul, 156
Sister's Ferry, 57, 58
Site national historique de la résistance en Vercors. See SNHRV
"Smalarina," 119
Smith, Dennis Mack, 127
Smith, Maj. Gen. Gustavus, 52
SNHRV (Site national historique de la résistance en Vercors), 150, 155–56, 157, 160, 162
Somme, Battle of the, 80, 82
Soper, Fred, 120, 121, 124, 125
South (United States), 25, 49
South Carolina, 56–63;
    topography of, 57
Southeast Asia, 2, 32, 41
Southeast Asian teak, 33
South Korea, 21

South Pacific, 112, 124
Soviet bloc, 187
Soviet territory, 187
Soviet Union, 1, 11, 20, 22, 104, 186
"space junk," 28
Spain, 11, 16, 23, 36, 38
Sparta, 62
"Spartacus in Green" (Leberecht Migge), 173, 174
spatiality of war, 68, 69, 74
Special Field Order No. 120, 50, 53, 57, 58–59, 62
Speer, Albert, 94, 176, 180
Spice Islands, 36
Spree River, 185
SS, 99
Stadt der Zukunft (Gartenstadt), Die (The City of the Future: Garden City), 179
Stadtlandschaft (cityscape), 178, 184, 188
stalemate war (World War I), 72, 75, 79, 82. *See also* entrenchment landscapes
Stalin, 172
Stalingrad, 92, 187
Stalinist urban reconstruction models, 187
Stations of the Cross, 158, 164
Stayer, Gen. Morrison, 120
Steffan, Hans, 180
Stellungskrieg (stalemate war), 75
Stephan, Hans, 181
Stewart, Mark, 49
Stone, Col. William, 120
Stuttgart, 95, 177–78
Subic Bay, 16, 22
Subic Bay Naval Base, 22
Sucesos de las Islas Filipinas, 38
Sudan, 1
Sulit, Carlos, 41, 48n71
Sultan Kudarat of Magindanao, 36
Sulu, 36
Sutter Bypass, 139, 147n22
Sutter National Wildlife Refuge, 139
Switzerland, 154

Tamms, Friedrich, 180
Tassigny, General de Lattre de, 150
Technical University of Berlin, 179

Tenza, Alonso Fajardo de, 38
"Test-Tube Island" (San José Island), 22
Teulada, 127
Third Reich, 181. *See also* Reich
Thucydides, 1
Tiber Delta, 112, 117, 118, 121, 124
timber, use of for military purposes, 1, 2, 4, 32–42
Tobacco Monopoly, 38–39
Todt, Fritz, 94
Tognotti, Eugenia, 126
Tokyo, 6, 19, 176–77, 187
Tokyo, 1923 Earthquake, 176
Toller, Ernst, 84
Torrejon (airfield), 23
total war, 77, 81, 92, 93, 94
total warfare, defined, 81–82
toxaphene, 144
Treaty of Westphalia, 36
trench building, 72–74
trench fever, 80
trench foot, 80
trench warfare, 4, 68–84
trenches, conditions in, 69, 72–75; impact of vermin on, 79–80
impact of weather on, 78–79
trenches, sounds and smells of, 75–77
trenches, topography of, 72
Trommelfeuer (drumfire), 76
tropical agriculture, 17
tropical forestry, 33, 134
tropical hardware, 33
tropical medicine, 17
tuberculosis, 112, 172
Tucker, Richard P., 3
Tule Lake, 138
Tuke Lake internment camp, 141, 145
Tule Lake Refuge, 140, 141, 144
Tumauini, 39, *40*
typhus, 19, 80, 112, 113, 119–20, 121

Uekötter, Frank, 5, 92–111, 195
Union Army, 5, 49–55, 57–58, 60–62, 63
United Nations, 5, 120, 127
United Nations' Food and Agriculture Organization (FAO), 126

Index

United States, acquisition of foreign territories, 16–17
United States Armed Forces. *See* United States military
United States Army Corps of Engineers, 3–4, 14
United States Bureau of Biological Survey, 133, 135, 142
United States Bureau of Reclamation, 135, 140, 141, 146
United States Chemical Warfare Service, 142
United States Civil War, 4, 11, 14, 56, 113. *See also* Sherman's 1864–65 Campaigns
United States Department of Agriculture, 144
United States Department of Defense, 23, 26
United States Department of Energy, 23
United States Department of the Interior, 140
United States Fish and Wildlife Service (FWS), 5, 6, 133–34, 137, 138–46;
    lessened authority of, 140–41, 145
    use of DDT by, 143–44
    impact of FWS-DDT on fish and wildlife, 143–45
    migratory birds conservation by, 133–34
    federal refuges of, 141–42
United States Forest Service, 133, 145
United States Interstate Highway System, 25
United States military:
    defeat of Amerindians, 13
    dominance of, 20, 28
    expansion in size of, 11–13, 20
    global expansion of, 16–17, 20–22, 27–28
    industrial progress caused by, 14–15
    infrastructure development by, 13–14
    principal purposes of, 11, 27–28
    weapons manufacturing by, 14
    transformation to Western U.S., 13
United States military, energy:
    consumption, 17–18, 28
    dependence on oil, 18, 23
United States military, environmental impact, 3–4, 10–28;
    on domestic ecology, 23
    on global ecology, 4, 21–23, 28
    indirect effects, 24–25
    from military maneuvers, 21–22
    during post World War II period, 26–28
United States National Park Service, 133
United States Navy, 2, 11, 14, 20;
    expansion of, 15–18
    impact on environment of, 15–16
United States, 19th century foreign policy, 15
University of Hamburg, 117
University of Heidelberg, 117
Upper Silesia, 99, 101, 102
urban historians, 171
urban landscapes, 171, 187
urban planning, as related to biosphere, 171
urban planning for post-WWII rebuilding, 6, 186–88
urban planning models, 171–76; 179–84;
    British town plans, 182
    civil defense aspects of, 176–78, 181
    Nazi planning models, 179–80
    new town models, 178–79
USSR, 25, 184

Valchevrière, 158, 162, 163
Vassieux-en-Vercors. *See* Vercors
Vassieux-en-Vercors cemetery, 158
VDI-Fachausschuß für Staubtechnik (German society of engineers), 101
vegetation loss, as caused by warfare, 22, 26, 77
Vercors, 6, 150, 159–60, 162–64;
    destruction to, by German assault, 153–55
    as land of freedom, 159–60, 166
    creation of memorials, 6, 150–51, 155
    as a "natural fortress," 153, 155, 156–58, 165
    topography of, 150, 156
Vercors, German assault on, 6, 150, 153–55
Vercors memorials:
    environmental narrative of, 150–51, 152–53
    drawing on nature for dramatic effect, 158, 160–62
    purposes of, 153, 155, 158, 165
    as source of tourism, 160, 164–65, 166
    impact of weather on, 152–53, 162–64, 166

Verdun, Battle of, 77
Vereenigde Oostindische Compagnie (VOC) (Dutch East India Company), 36
vermin, 79, 80, 112
Veterans, French (World War II), 6, 8. *See also* Pionniers du Vercors
Viana, Francisco Leandro de, 36–37
Vichy regime, 153, 182
Vidal y Soler, Sebastián, 34, 35
Vietnam War, 22, 26; environmental effects of, 2, 27
Villard-de-Lans, 158, 162
Virginia, 14, 57. *See also* Army of Northern Virginia
Vismann, Cornelia, 74
Volkskraft (life force), 174
VX nerve gas, 22

Wagner, Martin, 173–75, 181, 183
Wagner, Robert, 94
Wallace, Henry, 133
Wandervögel, 71
"war against insects," 145
War and Nature: Fighting Humans and Insects with Chemicals from World War I to Silent Spring, 2
"war gases," 142
war metaphors, use in mosquito eradication efforts, 122–24
War Relocation Authority (WRA), 140–41, 145–46
War Relocation Authority (WRA) camps, 140–41, 146
Warsaw, 186, 187
Washington (State), 26, 133, 134
Washington, DC, 17, 140
Waterfowl Management Areas, 139. *See also* refuges
Waterloo, 113
weaponry, 6, 10, 13, 21, 34, 36, 42, 81, 184
weather conditions, as wartime threat, 78–79
Wehrmacht, 92
Weigley, Russell, 59
Weserbergland mountains, 100
Weser River, 100

West Germany, 21, 187; reconstruction of, 181
Western Front, 4, 68–84
Western United States, 132–146
wildlife, impact of war upon, 132. *See also* game animals
wildlife conservation, 5, 132–46
Wilson, Robert, 5, 132–49, 195
Wilson, President Woodrow, 17
Wolters, Rudolf, 180
wood, importance of, 32, 133
wood, as "prize" of war, 43
World War I:
    death toll of, 82
    entrenchment landscapes of, 72–75
    role of environment in trench warfare, 70, 75–79
    pre-war expectations, 70–72, 95
    use of gases in, 80–81, 90n85
    threat of conditions to soldiers, 69–70, 73–75, 78–80
World War II:
    destruction to cities, 176, 184–85
    impact of, on U.S. conservation programs, 133–34, 135–40, 145–46
    as opportunity to rebuild cities, 6, 176, 182, 184
Wortmann, Wilhelm, 178
Wriezen, 180, 181
Wuppertal, 185
Württemberg, 97
Würzburg, 185
Wutach Gorge, 92, 93, 94, 104–6
Wutach Gorge project, 95, 104–5
Wutach River, 92–93, 105

yellow fever, 17
Young, James E., 151
Ypres, 18, 82
Ypres, Second Battle of, 80–81

Zaragoza (U.S. airfield), 23
Zeller, Thomas, 105
Zentral-Bauleitung der Waffen-SS (Auschwitz-Birkenau's building department), 99
Zweig, Arnold, 82